Special Metal-Alloy Coating and Catalysis

Special Metal-Alloy Coating and Catalysis

Editor

Wangping Wu

Basel • Beijing • Wuhan • Barcelona • Belgrade • Novi Sad • Cluj • Manchester

Editor
Wangping Wu
Electrochemistry and
Corrosion Laboratory, School
of Mechanical Engineering
and Rail Transit
Changzhou University
Changzhou, China

Editorial Office
MDPI
St. Alban-Anlage 66
4052 Basel, Switzerland

This is a reprint of articles from the Special Issue published online in the open access journal *Metals* (ISSN 2075-4701) (available at: https://www.mdpi.com/journal/metals/special_issues/Special_Metal_Alloy_Coating_and_Catalysis).

For citation purposes, cite each article independently as indicated on the article page online and as indicated below:

Lastname, A.A.; Lastname, B.B. Article Title. *Journal Name* **Year**, *Volume Number*, Page Range.

ISBN 978-3-0365-8926-8 (Hbk)
ISBN 978-3-0365-8927-5 (PDF)
doi.org/10.3390/books978-3-0365-8927-5

© 2023 by the authors. Articles in this book are Open Access and distributed under the Creative Commons Attribution (CC BY) license. The book as a whole is distributed by MDPI under the terms and conditions of the Creative Commons Attribution-NonCommercial-NoDerivs (CC BY-NC-ND) license.

Contents

About the Editor vii

Wangping Wu
Special Metal-Alloy Coating and Catalysis
Reprinted from: *Metals* **2023**, *13*, 1555, doi:10.3390/met13091555 1

Olga Lebedeva, Larisa Fishgoit, Andrey Knyazev, Dmitry Kultin and Leonid Kustov
Electrodeposition of Iron Triad Metal Coatings: Miles to Go
Reprinted from: *Metals* **2023**, *13*, 657, doi:10.3390/met13040657 5

Xiaodong Xu, Dingkai Xie, Jiaqi Huang, Kunming Liu, Guang He, Yi Zhang, et al.
Influence of Pretreatment Processes on Adhesion of Ni/Cu/Ni Multilayer on Polyetherimide
Resin Reinforced with Glass Fibers
Reprinted from: *Metals* **2022**, *12*, 1359, doi:10.3390/met12081359 25

Wenming Liu, Zhiqiang Xu, Hongmei Liu and Xuedong Liu
Cause Analysis and Solution of Premature Fracture of Suspension Rod in Metro Gear Box
Reprinted from: *Metals* **2022**, *12*, 1426, doi:10.3390/met12091426 41

Xinghai Shan, Mengqi Cong and Weining Lei
Effect of Cladding Current on Microstructure and Wear Resistance of High-Entropy
Powder-Cored Wire Coating
Reprinted from: *Metals* **2022**, *12*, 1718, doi:10.3390/met12101718 53

Jie Huang, Tao Chen, Daqing Huang and Tengzhou Xu
Study on the Effect of Pulse Waveform Parameters on Droplet Transition, Dynamic Behavior of
Weld Pool, and Weld Microstructure in P-GMAW
Reprinted from: *Metals* **2023**, *13*, 199, doi:10.3390/met13020199 65

Xianxian Ding, Yalin Lu, Jian Wang, Xingcheng Li and Dongshuai Zhou
Effect of Ce Content on the Microstructure and Mechanical Properties of Al-Cu-Li Alloy
Reprinted from: *Metals* **2023**, *13*, 253, doi:10.3390/met13020253 87

Weiwei Dong, Minshuai Dong, Danbo Qian, Jiankang Zhang and Shigen Zhu
Shear Transformation Zone and Its Correlation with Fracture Characteristics for Fe-Based
Amorphous Ribbons in Different Structural States
Reprinted from: *Metals* **2023**, *13*, 757, doi:10.3390/met13040757 101

Tengfei Meng, Hongjin Shi, Feng Ao, Peng Wang, Longyao Wang, Lan Wang, et al.
Study on Nitrogen-Doped Biomass Carbon-Based Composite Cobalt Selenide Heterojunction
and Its Electrocatalytic Performance
Reprinted from: *Metals* **2023**, *13*, 767, doi:10.3390/met13040767 113

Jozef Dobránsky, Miroslav Gombár, Patrik Fejko and Róbert Balint Bali
A Determination of the Influence of Technological Parameters on the Quality of the Created
Layer in the Process of Cataphoretic Coating
Reprinted from: *Metals* **2023**, *13*, 1080, doi:10.3390/met13061080 127

**Ralph Gruber, Tanja Denise Singewald, Thomas Maximilian Bruckner, Laura Hader-Kregl,
Martina Hafner, Heiko Groiss, et al.**
Investigation of Oxide Thickness on Technical Aluminium Alloys—A Comparison of
Characterization Methods
Reprinted from: *Metals* **2023**, *13*, 1322, doi:10.3390/met13071322 145

About the Editor

Wangping Wu

Wangping Wu received a PhD in materials processing engineering from the Nanjing University of Aeronautics and Astronautics, China, in 2013. He joined Tel Aviv University, Israel, as a postdoctoral fellow in October 2013, before joining the Hochschule Mittweida University of Applied Sciences, Mittweida, Germany, and Technische Universität, Chemnitz, Germany, in September 2019 as a visiting scholar with the support of the China Scholarship Council (CSC). He currently works as an associate professor at the School of Mechanical Engineering and Rail Transit, Changzhou University, China. His research focuses on the synthesis, characterization, and performance of films and coatings made of the noble metals and their alloys; mechanical failure analysis; and nanopowders dispersed in polymer and additive manufacturing. He has published over 100 papers in peer-reviewed international journals.

Editorial

Special Metal-Alloy Coating and Catalysis

Wangping Wu

School of Mechanical Engineering and Rail Transit, Changzhou University, Changzhou 213164, China; wwp3.14@163.com or wuwping@cczu.edu.cn

Special metal alloy coating is an important material technology that has a wide range of applications in many fields. It combines the characteristics of special alloy materials and surface coating technology to provide excellent properties and enhanced functions for materials. Catalysis is a process that increases the reaction rate without being consumed by intervening in the chemical reaction process. Catalysis can be applied to various chemical reactions, thereby increasing the reaction rate, improving product selectivity, and enhancing reaction efficiency. Special metal alloy coatings and catalytic technology have broad application prospects. They can improve material properties, increase reaction efficiency as well as product selectivity, and promote the development of environmentally friendly chemical processes. With the continuous advancement of science and technology in addition to the increasing demand for high-performance materials and efficient chemical reactions, special metal alloy coatings and catalytic technologies will continue to play an important role, bringing more benefits to applications in various fields.

This Special Issue will provide a snapshot of the state of the art in metal, alloy coating and films. These metals, alloy coatings and films can act as corrosion and oxidation resistance barriers as well as catalysts for water splitting, among other catalysis applications.

Ten contributions (nine articles and one review) have been published in this Special Issue, covering the research topics of the fracture of metal parts, progress in coating growth and characterization techniques, processing condition–structure–property relations, the development of novel lubricants and advanced simulations relevant to tribological contacts and metal manufacturing, and the structural, morphological, corrosion, oxidation, wear, and catalysis properties of thin films.

Lebedeva et al. [1] reviewed the possibilities and future perspectives of the electro-chemical deposition of bimetallic compositions and alloys containing Fe, Co, Ni, Cr, W, and Mo. In addition to deposition from aqueous (classical) solvents, the advantages and perspectives of electrochemical deposition from ionic liquids (ILs) and deep eutectic solvents (DES) are briefly discussed.

For the phenomenon of metal fracture, Liu et al. [2] explored the causes of the fracture of the gearbox boom, and found that the nature of suspension rod fracture belonged to fatigue fracture. Dong et al. [3] investigated the shear transformation zone and its correlation with the fracture characteristics of FeSiB amorphous alloy ribbons in different structural states.

Ding et al. [4] investigated the effects of Ce content on the microstructures and mechanical properties of Al-Cu-Li alloys, and provided an economical as well as convenient method for improving the properties of Al-Cu-Li alloys via the addition of Ce.

In terms of coating preparation, Xu et al. [5] successfully prepared Ni/Cu/Ni multilayer coatings on glass-fiber-reinforced PEI resin via sandblasting and the activation/acceleration of a two-step metallization process. The influence of acceleration on the appearance quality of metallization on the PEI substrate was studied, and, at the same time, the mechanism of acceleration was investigated and addressed. Dobránsky et al. [6] optimized the deposition process of electrophoretic paints, after which they then analyzed and evaluated the thickness of a cataphoresis layer formed on an aluminum substrate.

Citation: Wu, W. Special Metal-Alloy Coating and Catalysis. *Metals* **2023**, *13*, 1555. https://doi.org/10.3390/met13091555

Received: 1 August 2023
Revised: 12 August 2023
Accepted: 29 August 2023
Published: 4 September 2023

Copyright: © 2023 by the author. Licensee MDPI, Basel, Switzerland. This article is an open access article distributed under the terms and conditions of the Creative Commons Attribution (CC BY) license (https://creativecommons.org/licenses/by/4.0/).

Shan et al. [7] studied the effect of a tungsten arc melting current on the microstructure and wear resistance of FeCrMnCuNiSi$_1$ coatings prepared from high-entropy powder-cored wire, and studied the influence of a melting current on the wear resistance of the coatings.

Huang et al. [8] used an image processing program to extract the dynamic behavior characteristics of the droplet transition and the weld pool in high-speed photography. The influence of the current waveform on the arc pressure and the impact of the droplet were quantitatively analyzed with different parameters.

Meng et al. [9] reported a nitrogen-doped biomass carbon (1NC@3)-based composite cobalt selenide (CS) heterojunction, which was prepared via a solvothermal method using kelp as the raw material. Structural, morphological, and electrochemical analyses were conducted to evaluate its performance. The overpotential of the CS/1NC@3 catalyst in the OER process was 292 mV, with a Tafel slope of 98.71 mV·dec^{-1} at a current density of 10 mA·cm^{-2}. The presence of the biomass carbon substrate enhanced the charge transport speed of the OER process and promoted the OER process, confirmed by theoretical calculations. This study provided a promising strategy for the development of efficient electrocatalysts for OER applications.

Gruber et al. [10] used some advanced measurement techniques, such as transmission electron microscopy (TEM) and Auger electron as well as X-ray photoelectron spectroscopy (AES, XPS), to characterize the oxide layer of aluminium alloy surfaces. The results illustrated in detail the strengths and weaknesses of each measurement technique. The XPS technique was proven to be the most reliable method to reproducibly quantify the average oxide thickness.

This Special Issue was well supported by a diverse range of submissions. In this Special Issue, there are various topics relating to special metal alloy coating and catalysis. It shows the latest research results of the authors. However, many issues in this area of research have not yet been explored and the dissemination of these results should be continued. As a Guest Editor, I hope that the research results presented in this Special Issue will contribute to the further progression of research on special metal alloy coating and catalysis.

Finally, I would like to thank all of the reviewers for their input and efforts in producing this Special Issue, as well as the authors for the papers that they have prepared.

Conflicts of Interest: The author declares no conflict of interest.

References

1. Lebedeva, O.; Fishgoit, L.; Knyazev, A.; Kultin, D.; Kustov, L. Electrodeposition of Iron Triad Metal Coatings: Miles to Go. *Metals* **2023**, *13*, 657. [CrossRef]
2. Liu, W.; Xu, Z.; Liu, H.; Liu, X. Cause Analysis and Solution of Premature Fracture of Suspension Rod in Metro Gear Box. *Metals* **2022**, *12*, 1426. [CrossRef]
3. Dong, W.; Dong, M.; Qian, D.; Zhang, J.; Zhu, S. Shear Transformation Zone and Its Correlation with Fracture Characteristics for Fe-Based Amorphous Ribbons in Different Structural States. *Metals* **2023**, *13*, 757. [CrossRef]
4. Ding, X.; Lu, Y.; Wang, J.; Li, X.; Zhou, D. Effect of Ce Content on the Microstructure and Mechanical Properties of Al-Cu-Li Alloy. *Metals* **2023**, *13*, 253. [CrossRef]
5. Xu, X.; Xie, D.; Huang, J.; Liu, K.; He, G.; Zhang, Y.; Jiang, P.; Tang, L.; Wu, W. Influence of Pretreatment Processes on Adhesion of Ni/Cu/Ni Multilayer on Polyetherimide Resin Reinforced with Glass Fibers. *Metals* **2022**, *12*, 1359. [CrossRef]
6. Dobránsky, J.; Gombár, M.; Fejko, P.; Balint Bali, R. A Determination of the Influence of Technological Parameters on the Quality of the Created Layer in the Process of Cataphoretic Coating. *Metals* **2023**, *13*, 1080. [CrossRef]
7. Shan, X.; Cong, M.; Lei, W. Effect of Cladding Current on Microstructure and Wear Resistance of High-Entropy Powder-Cored Wire Coating. *Metals* **2022**, *12*, 1718. [CrossRef]
8. Huang, J.; Chen, T.; Huang, D.; Xu, T. Study on the Effect of Pulse Waveform Parameters on Droplet Transition, Dynamic Behavior of Weld Pool, and Weld Microstructure in P-GMAW. *Metals* **2023**, *13*, 199. [CrossRef]

9. Meng, T.; Shi, H.; Ao, F.; Wang, P.; Wang, L.; Wang, L.; Zhu, Y.; Lu, Y.; Zhao, Y. Study on Nitrogen-Doped Biomass Carbon-Based Composite Cobalt Selenide Heterojunction and Its Electrocatalytic Performance. *Metals* **2023**, *13*, 767. [CrossRef]
10. Gruber, R.; Singewald, T.; Bruckner, T.; Hader-Kregl, L.; Hafner, M.; Groiss, H.; Duchoslav, J.; Stifter, D. Investigation of Oxide Thickness on Technical Aluminium Alloys—A Comparison of Characterization Methods. *Metals* **2023**, *13*, 1322. [CrossRef]

Disclaimer/Publisher's Note: The statements, opinions and data contained in all publications are solely those of the individual author(s) and contributor(s) and not of MDPI and/or the editor(s). MDPI and/or the editor(s) disclaim responsibility for any injury to people or property resulting from any ideas, methods, instructions or products referred to in the content.

Review

Electrodeposition of Iron Triad Metal Coatings: Miles to Go

Olga Lebedeva [1], Larisa Fishgoit [1], Andrey Knyazev [1], Dmitry Kultin [1] and Leonid Kustov [1,2,3,*]

[1] Department of Chemistry, Lomonosov Moscow State University, 1-3 Leninskie Gory, Moscow 119991, Russia
[2] N.D. Zelinsky Institute of Organic Chemistry, Russian Academy of Sciences, 47 Leninsky Prospect, Moscow 119991, Russia
[3] Institute of Ecology and Engineering, National Science and Technology University "MISiS", 4 Leninsky Prospect, Moscow 119049, Russia
* Correspondence: lmkustov@mail.ru

Abstract: The possibilities and future perspectives of electrochemical deposition of bimetallic compositions and alloys containing Fe, Co, Ni, Cr, W, and Mo are reviewed. The synthesis of two- and three-component materials, as well as compositionally more complex alloys, is considered. The method of synthesizing of materials via electrodeposition from solutions containing metal ions and metalloids is one of the most promising approaches because it is fast, cheap, and it is possible to control the composition of the final product with good precision. Corrosion, catalytic and magnetic properties should be distinguished. Due to these properties, the range of applications for these alloys is very wide. The idea of a correlation between the magnetic and catalytic properties of the iron-triad metal alloys is considered. This should lead to a deeper understanding of the interplay of the properties of electrodeposited alloys. In addition to deposition from aqueous (classical) solvents, the advantages and perspectives of electrochemical deposition from ionic liquids (ILs) and deep eutectic solvents (DES) are briefly discussed. The successful use and development of this method of electrodeposition of alloys, which are quite difficult or impossible to synthesize in classical solvents, has been demonstrated and confirmed.

Keywords: electrodeposition; alloys; ionic liquids; deep eutectic solvents; coatings; functional material; nanotechnology

1. Introduction

Various methods, such as plasma spraying, vapor deposition, magnetron sputtering and laser cladding electrodeposition, have been successfully developed for the synthesis of functionally graded material coatings (FGMC) with the composition and mechanical properties gradually changing for the optimal coating thickness. The synthesis of materials by electrodeposition (EDP) from solutions containing metal ions and metalloids and subsequently recovering alloy components is currently considered one of the most advanced techniques because it is fast, cheap, and provides the possibility of precise control of the composition of the final product. The required mechanical, corrosion, magnetic, and catalytic properties of deposited coatings are provided by varying the synthesis parameters. Electrodeposition conditions (pulse parameters, current density, electrolyte composition) can be finely tuned to control the electro-crystallization process in order to produce gradually varying composition and micro/nanostructure of the coatings [1] or amorphous coatings [2]. The electrodeposition method has been commonly used for the synthesis of FGMC due to its ease, convenience and low cost. One may find a few drawbacks to the electrodeposition process: there is a high risk of hydrogen evolution, and the inclusion of basic compounds (hydroxides) in the alloy deposits makes them powdered, stressed, or exfoliated. Some of the electrodeposition drawbacks may stem from the difficulty of co-depositing multiple elements together.

While highlighting the ability to form a smooth and homogeneous FGMC, alloys were divided into two groups: two- and three- component systems based on the metals of the iron triad with chromium or with carbon and phosphorus [3–11], and alloys containing tungsten or molybdenum. The alloy composition is chosen on the basis of its possible applications. Alloys can be applied as FGMC [12], catalysts [13], elements of micro- and nanoelectromechanical devices [14], and electrodes-electrocatalysts in the water electrolysis process [15]. The unique acid resistance and microhardness of two-component alloys based on the iron triad with Mo and W makes them an improved alternative to chromium coatings [16,17].

Ionic liquids (ILs) have already proven themselves as systems that are promising in various fields of application [18–25]. Since ionic liquids consist almost entirely of "free" charge carriers—cations and anions—their application in electrochemistry is especially interesting. The uniqueness of ionic liquids is associated with their high electrochemical stability, relatively high electrical conductivity, and the absence of a measurable saturated vapor pressure. In some cases, ionic liquids demonstrate significant advantages over traditional electrolytes. Currently, ionic liquids with an exceptionally large electrochemical window (5–9 V) have appeared, and they can be used in a cyclic mode of operation without any loss of properties or destruction of their structure, and the wide electrochemical window allows the deposition of metals with very negative redox potentials. The prospects for using ionic liquids in the processes of electrodeposition of metals and alloys and in the processes of preparation of metal and alloy nanoparticles by electrochemical methods are attractive [23,24]. ILs are good solvents for both organic and inorganic materials. ILs can be aprotic, and thus problems with regard to hydrogen ions intrinsic to protic solvents can be eliminated [25].

The review [26] considers the electrodeposition of five groups of metals (ordinary, light, noble, rare earth, and others) and their alloys in ionic liquids and deep eutectic solvents (DESs). We did not mention the preparation of nickel-phosphorus alloys and FeCoNi ternary alloys, but the properties and preparation of alloys based on nickel and cobalt are briefly described.

From a chemical point of view, ILs and DESs are two separate groups of substances to be used in the preparation of alloys. DESs exhibit several advantages over ILs, such as their easy preparation and easy availability from relatively inexpensive components (the components themselves are well-characterized in terms of their toxicity, so they can be easily shipped for large-scale processing); they are, however, in general less chemically inert. The crucial difference between ILs and deep eutectic solvents is the wide variety of ionic species present in DESs, while ILs mainly consist of one discrete type of the anion and cation [27].

The objects of this review are presented in Figure 1.

A classification of metal co-deposition based on the thermodynamic approach is proposed by Brenner [28]. In the case of "normal" co-deposition, a more noble element (having a higher equilibrium potential value) is deposited more easily, and the composition of the precipitate corresponds to the composition of the solution. "Abnormal" co-deposition can be represented by "anomalous" and "induced" behavior. Anomalous co-deposition means that less noble metals are preferably deposited. Induced co-deposition indicates that a metal that cannot be deposited in its pure form can be co-deposited as an alloy.

The fundamental aspects of co-deposition from the point of view of kinetics were developed in the work of Landolt [29]. It was proposed that the measured current density at a mixed electrode is the sum of the partial current densities of all anodic and cathodic reactions. Three types of coupling of partial reactions can be distinguished in the case of alloy deposition: non-interactive co-deposition, transport coupled co-deposition, charge transfer coupled co-deposition. These three types of coupling behavior of partial reactions during co-deposition provide a useful qualitative description of the observed behavior.

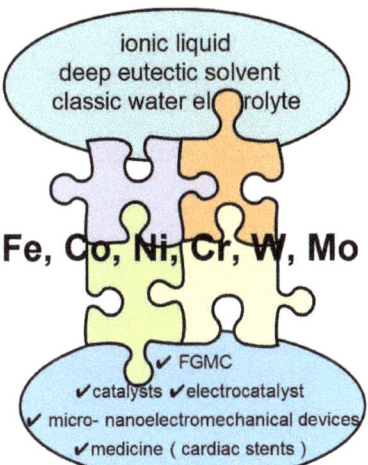

Figure 1. Bi- and ternary alloys of metals of the iron triad with chromium, molybdenum, and tungsten.

2. Electrodeposition from Water Solutions

2.1. Bi- and Ternary Alloys of Metals of the Iron Triad with Chromium and Metalloids

There are three main methods for EDP of alloys: the potentiostatic technique, direct current, and pulse current. For industrial applications, a two-electrode scheme is more suitable. For laboratory investigations, a three-electrode scheme is preferable. Schematic illustration of electroplating is presented in Figure 2.

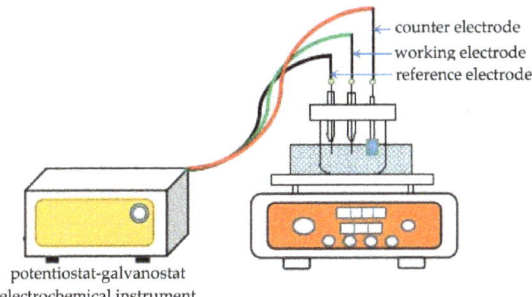

Figure 2. Schematic representation of the electrochemistry and electrodeposition experiment apparatus. Reproduced from [30] with permission from MDPI, 2023.

The most common electrodeposited materials are Ni–Fe alloy coatings [31], which, due to their remarkable electrocatalytic (reactions of hydrogen and oxygen production, CO_2 reduction), magnetic, and mechanical properties, have attracted attention in both scientific and industrial spheres. Their use in order to save Ni is economically more profitable.

Amorphous FeCr alloy films with a chromium content varying from 2.3% to 32.0 at% were deposited on a copper foil from an aqueous solution containing N,N-dimethylformamide by a potentiostatic electrodeposition technique [32]. The saturation magnetization for FeCr alloy films decreased but the microhardness increased with an increase in the chromium content.

The rate of Fe electrodeposition from the electrolytic bath is higher than that of Ni [33]. Ni-Fe coatings demonstrate unique magnetic properties.

Nanocrystalline Ni-Fe coatings were electrodeposited on steel substrates [3]. Coatings with a Fe content of 34.5% were obtained with a current efficiency of about 80%. It is shown that Ni^{2+} and Fe^{2+} are electrodeposited together to form a single Ni_3Fe phase, and Ni deposition is inhibited by the presence of Fe^{2+}, while Fe deposition is enhanced by the presence of Ni^{2+}. The nucleation and growth of nanocrystalline NiFe coatings is an instantaneous nucleation process controlled by 3D diffusion. The average grain size is 4.8 nm, and the roughness is less than 5 nm, which is better than that of a pure Ni coating. The NiFe alloy coating demonstrates good corrosion resistance (i_{corr} = 0.7544 µA cm^{-2}, R = 8560 Ω) due to the nanocrystalline compact surface.

The magnetic field was applied simultaneously to the process of electrodeposition of Ni-Co alloy with the Co content ranging within 21.80–59.55 at% from an aqueous solution [34]. The magnetoelectrodeposited alloys exhibited high corrosion resistance in comparison with a normal electrodeposited Ni–Co coating.

The effect of ultrasonic treatment on the thermal expansion, microstructure, and mechanical properties of electrodeposited FeNi layers was studied [4]. In order to avoid high transient cavitation energy, periodic ultrasound was introduced into the electrochemical process. Periodic ultrasound weakens the stripping effect as a result of the high ultrasound power, which allows one to use both a high current density and a high ultrasound power in the electroplating process. The iron content in the electrodeposited FeNi layer increased with increasing current density. The grain size decreased with increasing the number of cycles and growing current density. With a duty cycle of 0.57 and a current density of 1 A/dm^2, a highly efficient FeNi layer with excellent surface quality is obtained (roughness = 0.95 microns; iron content = 63.00 wt.%; microhardness = 373.1 NV; Young's modulus = 133.7 MPa; coefficient of thermal expansion = 5.4 × 10^{-6}/°C). Comparative characteristics of electrodeposited NiFe alloys are given in Table 1.

Table 1. Electroplating process parameters and some characteristics of deposits.

Parameter	Ref. [3]	Ref. [4]
Electroplating process parameters		
Substrate	steel	steel
Temperature (°C)	50	55
Cathode current density (A/dm^2)	3	1
Periodic ultrasound application	no	yes
Characteristics of deposits		
Roughness (nm)	5	95
Iron content (wt.%)	34.55	63.00
Grain size (nm)	4.8	11–12

The ultrasonic treatment resulted in a sharp growth of the grain size and an increase in the surface roughness factor (Table 1), which has mostly a negative effect on the properties of the alloy. The increase in the iron content in the electrodeposited alloy is presumably associated with a four-fold higher concentration of Fe^{2+} ions in the solution (20 g/L [3] and 80 g/L [4]).

Electrodeposition offers better control over the microstructure, shape, and composition of the deposit. Ternary NiFeCo and NiFeCr alloys were deposited in aqueous solutions [35]. Both alloys exhibit superior stability as compared with their lower-order alloy and pure metal samples, with the critical temperature for grain growth (or phase decomposition) improving by nearly 100 °C in the case of NiFeCo and by 200 °C in the case of NiFeCr. Electrodeposition is a viable route towards the synthesis of strong and highly stable nanocrystalline medium-entropy alloys (MEAs, containing 3–4 elements). The electrocatalytic activity in the hydrogen evolution reaction (HER) and oxygen evolution reaction (OER) in alkaline media for ternary nanostructured NiFeCo coatings and binary coatings electrodeposited in aqueous solutions was compared [36]. The comparison of the electrocatalytic activities of electrodeposited crystalline NiFeCo and amorphous NiFeCoP in HER showed the superior properties of NiFeCoP [37]. The authors [36,37] explain the high electrocatalytic activity by more developed surface and the synergy of the alloy components.

The electrochemical formation of Cr-C coatings was also studied [5]. The electrochemical formation of Cr-C coatings is realized through the simultaneous reduction of Cr(III) and decomposition of an organic ligand. The chemical state of metalloids (for example, carbon, phosphorous, and boron) in coatings electrodeposited from Cr(III) solutions may be determined with Valence-to-Core X-ray Emission spectra (vtc-XES, vtc-X-ray). The vtc X–ray spectra allowed one to determine the presence of metalloid atoms that are covalently bound to metal atoms and to estimate their quantitative content in metal-metalloid coatings formed by various methods (Figure 3).

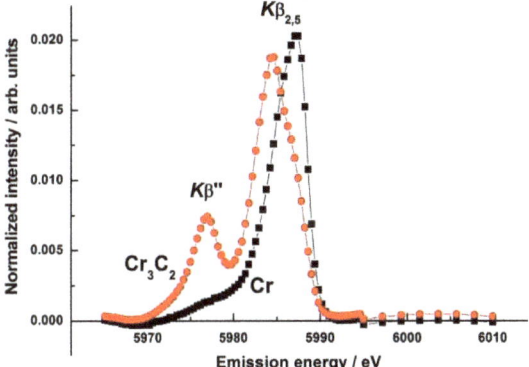

Figure 3. Normalized vtc X–ray emission spectra of metallic chromium and Cr3C2 carbide used in modeling experimental spectra of vtc X-ray of Cr–C samples. Reproduced from [5] with permission from Elsevier, 2018.

The comparison of vtc X–ray data with data on the total amount of carbon in the samples obtained, for example, by X-ray diffraction makes it possible to divide the total amount of carbon in the samples by the amount of elemental carbon and carbon that is covalently bound to chromium atoms. It should also be noted that for the obtained coatings with a carbon content below 40 at.%, an increase in the size of crystallites was observed after calcination at 500 °C. Coatings with a high carbon content do not show any signs of long-range order in their crystal structure.

Electrodeposition of metal–metalloid alloys of the NiP type was considered [6,7]. The effect of heat treatment and variation of the phosphorus content on the magnetic properties of electrodeposited NiP alloys was investigated [6]. The magnetic properties of the alloys obtained were explained with the use of differential scanning calorimetry (DSC) data. Magnetization measurements and DSC analysis of NiP alloys showed that the alloys with a phosphorus content exceeding 12 at.% were paramagnetic due to the absence of exchange interaction as a result of the formation of a network of P–rich paramagnetic domains. Amorphous NiP alloys, which were originally paramagnetic, became ferromagnetic after the heat treatment, which also led to their devitrification. The transition to the ferromagnetic state occurred as a result of the formation of the ferromagnetic phase of nickel, while the coercive force of the alloy increased due to an increase in the crystallite size and an increase in the proportion of the paramagnetic phase (Figure 4).

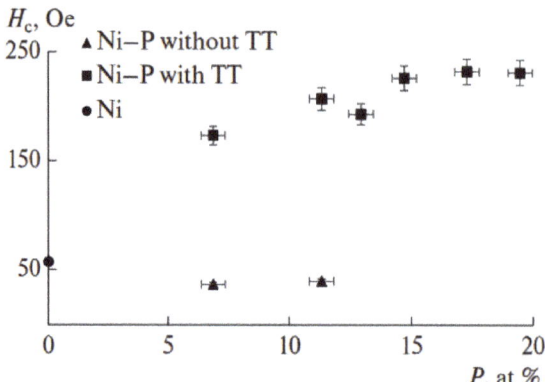

Figure 4. Effect of the phosphorus content on the coercive force of Ni–P coatings in the initial state and after thermal treatment relative to the coercive force of pure nickel. Reproduced from [7] with permission from Pleades Publ., 2017.

Regardless of the heat treatment, the magnetization of the alloys and the saturation magnetization decreased with increasing phosphorus content in the alloy.

Electrodeposition of Ni-P coatings was performed from an aqueous solution by direct current using a stirring time-controlled technique. Stirring leads to the formation of Ni-P layers with a higher P content, while without stirring Ni-P layers with a lower P content are produced [38].

The composition of NiP alloys depends on the content of sodium hypophosphite in the electrolyte in the course of the electrochemical deposition of an amorphous nickel coating. It was found for the first time [7] that alloys obtained under similar deposition conditions from electrolytes with fixed concentrations of nickel sulfate and sodium citrate have phosphorus concentrations ranging from 0 to 22–23 at.%. The Curie temperature of the obtained Ni-P samples, which are ferromagnetic in their initial state ([P] < 12 at.%), is in the range of 150–200 °C. At the same time, all heat-treated Ni-P samples demonstrated a Curie temperature close to that of pure nickel in the range of 350–360°C. The initial electrodeposited alloys become paramagnetic when they contain phosphorus in concentrations higher than 12 at.%. It is due to the lack of the possibility of exchange interaction as a result of fluctuations in the chemical composition. Paramagnetic alloys of the NiP type, after heat treatment, which leads to their uncovering, exhibit ferromagnetic properties due to the release of the ferromagnetic phase of Ni. An increase in the phosphorus concentration leads to an increase in the coercive force as a result of an increase in the grain size and in the proportion of the paramagnetic phase. The residual magnetization and saturation magnetization of alloys decrease with the increasing phosphorus concentration in the case of both heat-treated and non-heat-treated alloys.

The results of these works [6,7] could serve as a basis for theoretical predictions and systematic and rational analysis of various factors affecting the magnetic properties of materials containing several ferromagnetic elements.

The mechanism of electrochemical deposition of NiCrP alloys in a glycine bath was studied [8]. Electrodeposition on a copper plate was carried out at a current density of 10–20 A/dm^2. The electroreduction mechanisms of individual Cr and Ni, NiCr, NiP, CrP, and NiCrP alloys were studied. The possible process of trivalent chromium reduction can be explained by the formation of a glycine complex of bivalent chromium $[Cr(H_2O)_4(Gly)]^+$ (Equation (1)). The bivalent chromium complex was directly reduced to metallic chromium (Equation (2)). The reduction process of nickel is described by Equation (3). The key to NiCr co-deposition is the ΔE value for the reduction of nickel and chromium ions. During electrodeposition of the NiCr alloy, nickel is initially deposited on the surface of the cathode (Equation (3)). Nickel acts as a catalyst, causing a significant positive shift in the initial reduction potential of Cr(II) ions, thus satisfying the potential difference (approximately −180 mV) for the co–deposition of NiCr.

Nevertheless, the actual potential difference between Ni and Cr is the same as in the case of NiCr co–deposition, which ensures electrodeposition of the three–component alloy for the NiCrP system. In addition, it has been demonstrated that P tends to co-precipitate with Cr during the electrodeposition of NiCrP, which can be explained by the difference in the behavior during the electrodeposition of NiP and CrP alloys. The possible reduction processes can be presented as follows:

$$[Cr(H_2O)_4(Gly)]^{2+} + 1\,e = [Cr(H_2O)_4(Gly)]^+ \tag{1}$$

$$[Cr(H_2O)_4(Gly)]^+ + 1\,e = Cr^0 + Gly + 4H_2O \tag{2}$$

$$Ni^{2+} + Ni^+{}_{ads} + 2\,e = Ni^0 + Ni^+{}_{ads} \tag{3}$$

The possible mechanism of Ni–P electrodeposition can be explained as follows (Equations (4) and (5)):

$$H_2PO_2^- + 5H^+ + 4\,e = PH_3 + 2H_2O \tag{4}$$

$$2PH_3 + 3Ni^{2+} = 3Ni^0 + 2P + 6H^+ \tag{5}$$

The inorganic complexing agent used to obtain a CrP alloy is $NaH_2PO_2 \cdot H_2O$. The P incorporation mechanism in the course of the CrP alloy deposition is considered to be different from that realized in the case of the NiP alloy deposition.

The techniques used to produce multilayer and gradient coatings have been demonstrated in a review [39]. Ni-W multilayers were produced by varying the direct current density. Comparison of the direct and pulse currents on the properties of deposits was performed. The technique controlling the electrode potential is also effective for multilayer coating production. Metals of the Fe-triad are commonly used in multilayer coatings electrodeposited by controlling the electrode potential. The necessity of studying the electrode kinetics and the effects of various additives is shown.

So, a trivalent chromium solution containing glycine and $NaH_2PO_2 \cdot H_2O$ may also form a complex with hypophosphite, i.e., the $[Cr(H_2PO_2)(H_2O)_5]^{2+}$ complex. Taking this into account, the positive shift in the onset potential of CrP codeposition is associated with the increase in the concentration of electroactive Cr(III) complexes near the cathode (Equations (6)–(8)):

$$[Cr(H_2PO_2)_4(H_2O)_5]^{2+} + 1\,e = [Cr(H_2PO_2)_4(H_2O)_5]^+ \tag{6}$$

$$[Cr(H_2PO_2)_4(H_2O)_5]^+ + 2\,e = Cr^0 + 5H_2O + H_2PO_2^- \tag{7}$$

$$3H_2PO_2^- = H_2PO_3^- + 2P + H_2O + 2OH^- \tag{8}$$

The XRD analysis showed that the electrodeposited alloy coatings had an amorphous structure. The incorporation of phosphorus in the Ni deposit generally increases the number of defects in the crystalline lattice of electrodeposited alloy coatings, thereby transforming the coatings from a crystalline to an amorphous state. Scanning electron microscope images showed that a smooth and compact NiCrP coating was obtained at a current density below 15 A/dm^2 (Figure 5).

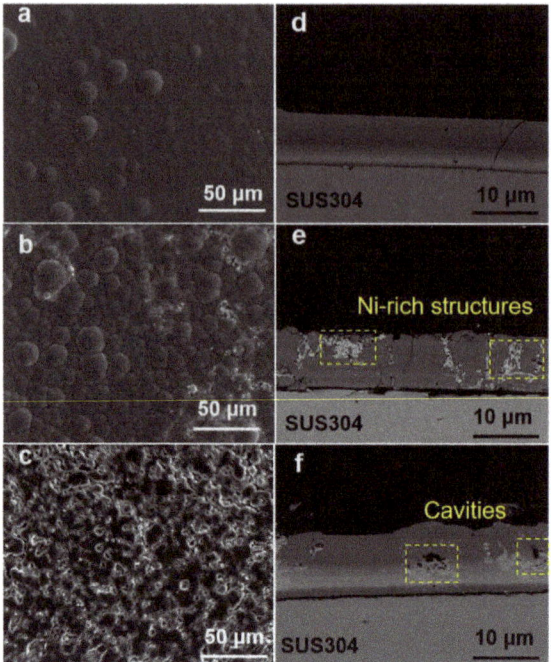

Figure 5. SEM micrographs of the surface (**a–c**) and cross-sectional micrographs (**d–f**) of coatings prepared from electrodeposited NiCrP alloys at the current densities (**a,d**) 10 A/dm^2, (**b,e**) 15 A/dm^2, (**c,f**) 20 A/dm^2. Reproduced from [8] with permission from Elsevier, 2021.

Electrodeposited NiCoP, FeCoP, and FeNiP alloys have been prepared [9–11]. The authors [9] applied magnetic assisted jet electrodeposition for effectively changing the microstructure and properties of NiCoP alloy films. The advantage of this method is inhibition of the growth of pores on the surface of the film and improving the adhesion of the coating to the surface. An increase in the saturation magnetization of the coating was observed as compared with simple jet electrodeposition.

The effect of calcination on the structure of coatings is analyzed using a combination of vtc-XES and X-ray diffraction to conclude whether recrystallization involves any redistribution of covalently bound metalloid between atoms of different metals in a three-component system [10,11]. The NiCoP, FeCoP, and FeNiP alloys were electrodeposited on a copper substrate from solutions containing Fe(II), Co(II), and Ni(II) with NaH$_2$PO$_2$ additives were taken as the object of the study. The FeNiP system was described in detail in ref. [10].

The corrosive and mechanical properties of cobalt-nickel-phosphorus ternary alloy coatings are discussed [40]. The role of pH, bath composition, and conditions of electrodeposition are summarized.

The vtc-XES data show that the coatings contain a high concentration of chemically bound phosphorus. Comparison of the vtc-XES spectra of all initial and calcined coatings [10,11] allows one to unequivocally conclude that the concentration of chemically bound phosphorus changes slightly due to crystallization, as for the two-phase electrodeposited alloys described above [6,7].

2.2. Iron Triad Metal Alloys with Molybdenum and Tungsten

Nickel-molybdenum alloys are promising electrocatalysts for HER. There are two main methods for EDP of NiMo alloys, namely, electrodeposition by direct current and by the pulsed current [40–43]. The method of EDP by pulsed current is preferable since it allows one to apply also the duty cycle or effective time of the applied current.

The relationship between independent variables and coating properties was established. In particular, the effect of nickel sulfate and sodium/molybdenum concentrations on various electroplating reactions was evaluated. Coatings with a high percentage of Mo and distinct morphology were obtained. X-ray diffractograms showed that all the samples were amorphous. The optimal bath composition was found at the ratio Ni:Mo of 7.5:5. A film with a maximum Mo content of 29 wt.% was obtained. However, the alloy with the best corrosion properties was deposited from a bath with a Ni:Mo component ratio of 10:3. The NiMo alloys were deposited by the pulsed current [42]. The morphology of the samples and the Mo content in the alloys were affected by a change in the working cycle from 70% to 30%. The material exhibited a crystalline structure. All the analyzed NiMo alloys showed activation of the hydrogen release reaction at a 30% duty cycle. The mechanism of the electrodeposition was proposed based on the published results [44–47].

The corrosive, mechanical, and magnetic properties of alloys containing metals of the iron group with tungsten are highlighted in a review [48]. Modern codeposition mechanisms are critically discussed.

The two-component alloys of the NiW and CoMo types were electrodeposited on a Cu substrate by the direct current method [49]. The deposition of the CoMo composition was carried out at $i = 30–120$ mA/cm^2, $t = 10–60$ min, $T = 298$ K, pH = 5.0–9.0. The alloy composition (wt.%) was determined by EDX: O(2–10), Co(13–87), Cu(0.7–7.6), Mo(7–50). All samples are characterized by a small coating thickness (less than 30 microns) and by the presence of copper in the alloy, which makes them translucent from the substrate. The presence of oxygen (up to 10 wt.%) in the composition of the electrodeposited coating indicates the products of incomplete reduction, i.e., oxide phases. The surface quality of alloys with a sufficient amount of Mo is bad. Cracks can be seen as a result of accumulated micro stresses and possible formation of hydrogen clusters, also described elsewhere [50,51]. Summing up the mechanism of the joint electrodeposition of cobalt and molybdenum according to the results available in the literature [46,52,53], it can be stated in the simplest version as follows: initially, metallic cobalt is electrodeposited on the cathode, a thin layer of which acts as a catalyst for the reduction of molybdenum ions with hydrogen. The formation of a joint complex of molybdenum and a precipitating metal contributes to the transfer of charges to the molybdenum ion through the precipitating metal. Molybdenum co-precipitates with metals of the iron triad due to the energy gain caused by alloying. The electrolyte containing sodium pyrophosphate as a complexing agent turned out to be more effective than a citrate-containing agent in terms of forming thicker coatings, however, no sample had sufficient thickness to study it by X–ray phase analysis, therefore, the approximate phase composition of the samples can be judged by the nickel-molybdenum state diagram. The deposits are basically formed in the two-phase region $Mo_6Co_7^+MoCo_3$. The deposition of the NiW system was carried out at $I = 80–300$ mA/cm^2, $t = 30$ min, $T = 298$ K. The alloy composition (wt.%) was determined by EDX: O(5–13), Ni(13–85), Cu(6–18), W(3–15). The deposits obtained at the maximum cathode current density demonstrated relatively low oxygen contents. The copper content in the alloy decreases with an increase in the cathode current density. The nickel content in the alloys increases with an increase in the current density.

The content of tungsten in the coatings increases with an increase in the concentration of tungstate ions in the solution and does not depend on the current density. The optimal composition is achieved in alloys obtained from the most concentrated solutions of tungstic acid salts at maximum current densities. Authors [51] proposed that non-electrochemical deposition of tungsten in the form of oxides was observed.

According to the Brenner classification, deposition of nickel-tungsten alloys is conducted in the form of "induced co-deposition". The joint electrodeposition of tungsten and nickel is divided into four main stages [54–57]:

1. Electrochemical generation of reactive forms of nickel and tungsten. The intermediate forms of these metals with an unpaired electron, i.e., particles of the radical type, exhibit a particularly high reactivity;

2. Formation of refractory metals from reactive particles of heteropoly compounds due to electrode initiation of the polymerization process. The film formed in this case has a low electron conductivity;
3. Electrochemical reduction of metal ions in the non-metallic system at the point of contact of parts of the film with ions in different oxidized states;
4. Final electrochemical reduction of metal ions at the film-alloy interface.

The alloy formation model will be valid if the film has low electron conductivity. Only in this case is a high negative potential achieved at the alloy-film boundary, and the release of hydrogen is blocked since the removal of gaseous hydrogen is very difficult. Consequently, the alloy is obtained at the film-alloy boundary by the following total reaction [55]:

$$[(Ni)(WO_4)(Cit)(H)]^{2-} + 8e- + 3H_2O \rightarrow NiW + 7(OH)^- + Cit^{3-} \qquad (9)$$

The presence of copper in deposits indicates a thin coating, and the excess of oxygen can be explained by the fact that the deposition of tungsten occurs in the form of tungsten blue $WO_{3n}(OH)_n$ ($0.1 \leq n \leq 0.5$).

Ternary alloys CoNiW were electrodeposited on a copper substrate by using a direct or pulsed current were studied [50,51,58,59]. The content of each element in the coating was in the range of 5–35 at.%, which corresponded to the definition of a high-entropy alloy [50]. A crack-free coating can be obtained by applying a pulsed current with an average current density of 15 mA/cm^2. All coatings obtained under various electrodeposition conditions had an amorphous structure, which indicated the formation of a solid solution on the substrate. Cracks, usually caused by internal stress or hydrogen embrittlement, can be observed on coatings obtained by the direct current method. The corrosion potentials of the coatings were lower than those of the substrate, but the corrosion current density decreased. The coating can increase the Cu substrate protection efficiency up to 73.8%. The influence of the bath composition and electrodeposition parameters on the structure, composition, surface characteristics and corrosion properties of NiWCo alloys was investigated [51]. Corrosion and wear characteristics were evaluated. It has been proved that the CoNiW alloy has a homogeneous, compact and flat surface exhibiting a colony-like morphology. An amorphous or nanocrystalline structure was formed depending on the process parameters, including the current density, pH, and electrolyte composition. The size of the crystallites did not depend notably on the Co content, but strongly depended on the current, pH of the solution and the synthesis time. The average roughness was optimized to 4–7 nm. The triple alloy contains 33 wt.% W, 21–60 wt.% Ni, and 2.8–40 wt.% Co. A decrease in the nickel content in deposits corresponds to an increase in the Co content. The W content somewhat changed while varying the electrodeposition conditions. The intermediates for the formation of the CoNiW alloy are $[Ni(HWO_4)(C_6H_5O_7)]^{2-}$ and $[Co(HWO_4)(C_6H_5O_7)]^{2-}$ ions. The addition of alumina nanoparticles slightly increases microhardness, reduces adhesive and oxidative wear, and significantly increases wear resistance [59]. The way that coatings wear out depends on the conditions of deposition. Alloys of the CoNiW type with a composition and structure gradually changing in thickness were synthesized using direct current electrodeposition [58–60]. A simple and convenient method of synthesis by induced electrodeposition is applicable to obtain multicomponent functional coatings [58].

Solution plasma sputtering is a simple and facile technique for the synthesis of NPs in a solution [61]. A schematic diagram demonstrates the formation of nanoparticles via the plasma sputtering method (Figure 6).

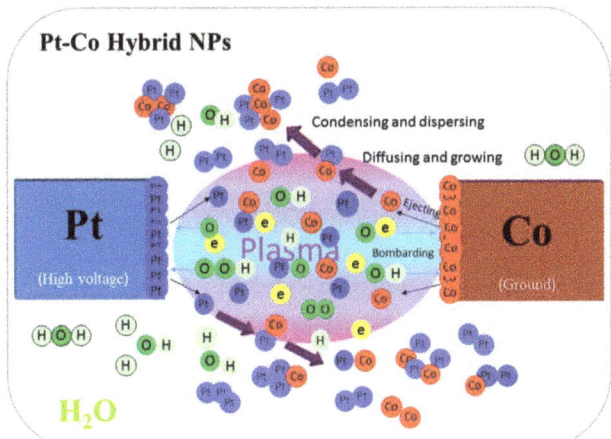

Figure 6. Schematic diagram for the synthesis of Pt/CoPt-1 composite NPs by solution plasma sputtering. Reproduced from [61] with permission from Nature, 2017.

The plasma generation technique in the liquid phase was used to synthesize alloy NPs with unique properties suitable for many applications [62].

3. Electrodeposition of Two- and Three-Component Alloys in Ionic Liquids and Deep Eutectic Solvents

3.1. Electrodeposition of Alloys in Ionic Liquids

The comparison of aqueous solutions and organic solvents with ionic liquids (ILs) as electrolytes is presented in the literature [63,64]. Some ionic liquids are manufactured on a commercial scale. The use of ILs in the electrodeposition of metals and alloys results in enhanced current efficiencies (CE > 90%) and production of corrosion-resistant and non-flaking coatings. The use of ILs in electrodeposition may be realized by three manners:

i. "pure" ILs
ii. ILs with additives [65]
iii. ILs as additives [66]

As received "pure" ILs may contain different impurities, mostly water. Such substances as acetonitrile, coumarin, thiourea, benzotriazole, acetone etc., are commonly considered as additives. By replacing some of the bulky adsorbed IL cations, the additive molecules can induce a more facile electrode reaction. As a consequence, a smoother and shinier surface is obtained in the presence of additives [65,67].

The electrodeposition of a Ni-Fe alloy from DES with water additives was studied [68]. The authors found that "abnormal" co-deposition occurs in the presence of water.

Ni-Co alloys were deposited from aqueous solutions with different metal salts and with additives of ILs 1-methyl-3-(2-oxo-2-((2,4,5-trifluorophenyl)amino)ethyl)-1H-imidazol-3-ium iodide ([MOFIM]I) and 1-(4-fluorobenzyl)-3-(4-phenoxybutyl)imidazol-3-ium bromide ([FPIM]Br) [66]. It was shown that the composition of alloys depended on the composition of the electrolyte (Table 2).

Table 2. The content of Co^{2+} in the electrolytic solution (wt.%) and the cobalt content (at.%) in the deposited Ni-Co alloys without and with additives.

Composition	Cobalt Content		
Composition of the electrolytic solution	30	50	70
Composition of deposited alloy			
1. without ILs	38.03	59.8	79.0
2. with [MOFIM]I	35.9	65.65.	85.33
3. with ([FPIM]Br	48.96	60.76	83.62

Electrodeposition of a Ni-Co alloy onto a Cu substrate was performed using an acidic sulfate bath in the absence and presence of different concentrations of [MOFIM]I and [FPIM]Br [19]. The corrosion behavior of coelectrodeposited Ni-Co alloys was performed in a marine water medium. A higher extent of inhibition of Ni^{2+} and Co^{2+} reduction is indicated by the increasing shift of the cathodic polarization curves towards more negative potentials (Figure 7).

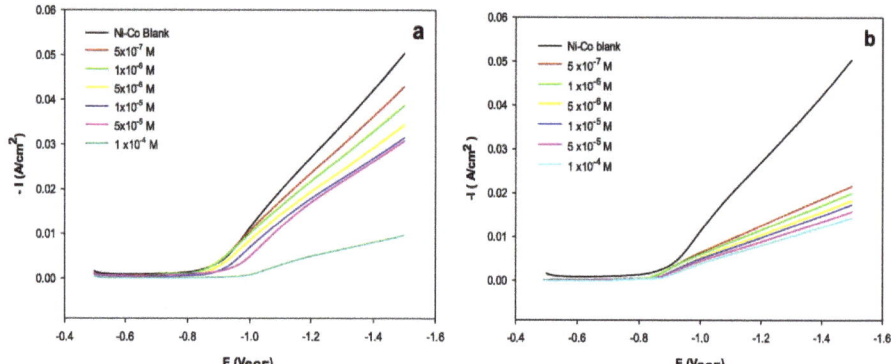

Figure 7. Potentiodynamic cathodic polarization curves for the Ni-Co alloy electrodeposition from the Ni70%-Co30% bath in the absence and presence of different concentrations of (**a**) [MOFIM]I, (**b**) [FPIM]Br at pH 4.5. Reproduced from [19] with permission from Elsevier, 2021.

The inhibition effect of the [MOFIM]I and [FPIM]Br molecules due to their adsorption on the cathode surface obeys the Langmuir adsorption isotherm. Compared with [FPIM]Br, [MOFIM]I reveals a better corrosion inhibition efficiency and more efficient additive properties. According to the natural atomic charges, [MOFIM]I demonstrates the strongest adsorption ability on the Cu substrate and a higher $IE_Rct\%$ compared with [FPIM]Br. These results show that the theoretical results are consistent with the experimental findings.

The study [20] shows the possibility of electroplating the surfaces of Fe, Ni, and Ni-Fe in ionic liquids as solvents without hydrogen evolution. It is important to remark that Ni-Fe alloy electrodeposition was successful in 1-butyl-1methylpyrrolidinium bis trifluoromethylsulfonyl)imide ([P1,4][Tf2N]) even though iron films are difficult to plate alone. An unexpected change of the alloy composition versus polarization (increase, decrease and further increase in the iron atomic percentage) was observed.

A summary of the recent literature discussing the use of ILs in the preparation of electrocatalysts based on Ni-alloys for OER and HER was presented [21]. It was shown in the review how ILs function as solvents and electrolytes for high-temperature electrodeposition baths, as well as structure-directing agents, doping agents, stabilizers, and/or capping agents in nanoparticle synthesis. For example, the mechanism for the synthesis of Ni_2P nanoparticles from the tetrabutylphosphonium chloride ([P4444]Cl) ionic liquid was proposed (Figure 8).

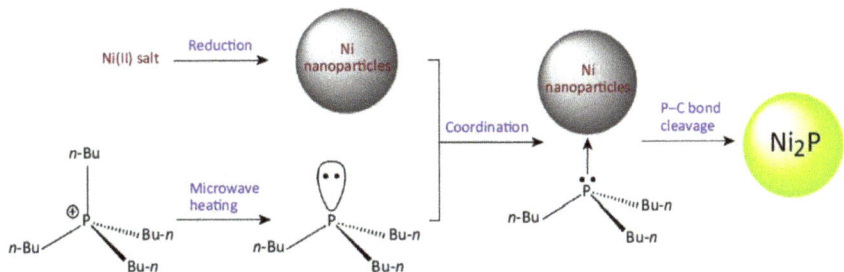

Figure 8. The mechanism for the synthesis of Ni2P nanoparticles from the [P4444]Cl ionic liquid. Reproduced from [19] with permission from Elsevier, 2021.

The IL was found to be the source of phosphorus in the resulting nickel phosphides through a suggested mechanism involving decomposition under heating to generate tributylphosphine, coordination of tributylphosphine with Ni(0) nanoparticles, and subsequent cleavage of the remaining P–C bonds as shown in Figure 8 (i.e., capping and doping). Understanding the role of the structure of the double electric layer of IL in electrochemical processes, in particular in the electrodeposition of metals and alloys, will allow one to predict the conditions for obtaining deposits with the desired properties [69].

So, the use of IL in industry is still at the very early stage [64]. The authors [21] stressed that often it is not obvious why a given IL is selected for a chosen task. The major drivers for industrial use of ILs are their cost and availability. The other reason for the restriction on the commercialization of some ILs is their toxicity (imidazolium-based ILs). The major limitations for industrial electrodeposition of metals and alloys in ILs are [63]:

- The relationship between the precipitate structure and the composition of the IL has not been studied in detail yet;
- Coatings must achieve quality standards, and process development is required to a large extent;
- Some applications are at the fundamental research stage with associated higher risk, that is, electroless, semiconductor, anodizing, and nanocomposite coatings;
- Process economics has been determined for a limited number of processes.

Only when these problems are properly solved will the full potential of these unique solvents in the production of efficient electrodeposited electrocatalytic materials be realized.

3.2. Electrodeposition in Deep Eutectic Solvents (DESs)

DESs based on choline chloride (ChCl) and other quaternary ammonium salts are most often applied for the deposition of metals and alloys [15,27]. Examples of compounds forming deep eutectic solvents (DES) with ChCl are anhydrous and crystal hydrates of metal salts, urea, ethylene glycol (EG), and many others. Chromium precipitation from aqueous solutions occurs with a rather low current efficiency and is accompanied by precipitation of basic compounds (salts, hydroxides). The inclusion of the latter in the precipitate led to the formation of powdered, strained, or flaking precipitates [70]. DESs usage let one to obtain dense precipitates with greater adhesion.

According to Abbot [71], DES is promising for obtaining coatings with new compositions and unique properties. The review discusses the effect of additives (especially water) on the electrodeposition of metals and alloys from DES. Water changes the structure of the double electric layer and improves mass transfer [71]. The electrodeposition of a Ni-Fe alloy from DES with water additives (up to 15 wt%) was studied [68]. The authors found that "abnormal" co-deposition occurs in the presence of water. The review by Smith [72] noted that, despite the fact that DES is economically difficult to compete with aqueous electrolytes, they may be of some value for industrial applications.

Nanoscale Fe-Cr alloy was successfully electrodeposited in a choline chloride-ethylene glycol (ChCl-EG, mole ratio 1:3) deep eutectic solvent in a two-electrode system by optimizing the concentration ratio $C_{Fe(II)}/C_{Cr(III)}$ [18]. ESI-MS analyses indicate that $[Fe(H2O)_3(Cl)_3]^-$ and $[Cr(H_2O)_2Cl_4]^-$ are the dominant species of Fe(II) and Cr(III) in ChCl-EG DES, respectively. Linear sweep voltammetry demonstrates that with the increase in $C_{Fe(II)}/C_{Cr(III)}$ from 1:5 to 1:1, the reduction potential difference between Fe(II) and Cr(III) becomes smaller, which is conducive to the electrodeposition of a Fe-Cr alloy deposit with a higher Cr content. The reduction of Fe(II) or Cr(III) on a glassy carbon electrode is a quasi-reversible process controlled by diffusion, with diffusion coefficients of $5.34 \cdot 10^6$ cm^2/s and $2.22 \cdot 10^6$ cm^2/s, respectively. FE-SEM observation shows that as the $C_{Fe(II)}/C_{Cr(III)}$ ratio decreases from 1:2 to 1:5, the microstructure becomes non-uniform, and the morphology transforms from homogeneous particles to scaly blocks. At the ratio of 1:2, the prepared nanocrystalline Fe-Cr alloy exhibits a symmetrical element distribution with a mean coating thickness of 120.3 µm and a mean diameter of 1.56 nm (Figure 9). Good corrosion resistance was revealed for the prepared Fe-Cr alloy. All the above studies provide a theoretical foundation for Fe-Cr alloy production by varying the electrolyte ratios.

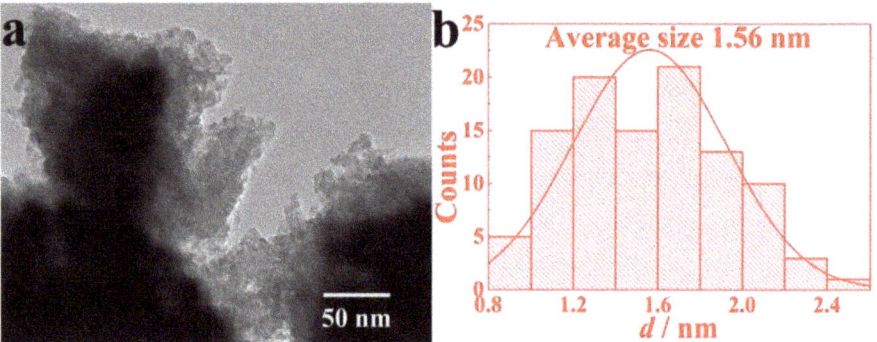

Figure 9. Fe-Cr alloy synthesized at the concentration ratio $c_{Fe(II)}/c_{Cr(III)}$ of 1:2 in ChCl-EG DES. (**a**) TEM image; (**b**) The particle size histogram. T = 333 K; U = 2.8 V. Reproduced from [18] with permission from Elsevier, 2022.

Ni-Co coating was deposited from DES by a potentiostatic technique on glassy carbon and Si/Ti/Au electrodes [73]. The CoNi films presented two CoNi alloyed crystal structures; therefore, the formation of Ni cubic crystalline structures and amorphous Co nanocrystalline structures was promoted. The deposits were tested as platforms to activate the formation of sulfate radicals.

The effects of the deposition potential on the morphology, chemical composition, crystal structure, corrosion resistance, and magnetic properties of nanostructured Fe-Co-Ni coatings were investigated [22]. The Fe-Co-Ni coatings prepared from ChCl/2Urea DES at different deposition potentials showed good corrosion resistance. Fe-Co-Ni coating that is produced at the deposition potential of −1.2 V (vs. saturated calomel reference electrode), exhibited a more positive corrosion potential and obvious magnetic anisotropy. The effective electrodeposition of nanostructured Co-films with a high surface area was performed by DES [74].

The coatings with the Fe contents of 89, 69, 47, and 28 at% were electrodeposited from DES [75]. The electrocatalyst with 69% Fe demonstrated high activity in HRR in alkaline media. The comparison of the alloy and the bath composition showed a close match.

The Fe and Fe-Ni coatings were electrodeposited in novel DES consisted of mixtures of FeCl$_3$ (or NiCl$_2$) and acetamide [76]. The Fe–Ni alloys demonstrated high corrosion resistance. The 1:1:10 FeCl$_3$–NiCl$_2$–acetamide mixture produced three deposits with compositions: Fe and alloys Fe72Ni28 and Fe12Ni88.

NiMo alloys were obtained from DES by the pulsed technique [42] at potentials ranging from −0.5 to −0.9 V vs. Ag [77] and with the conventional Watts bath [78].

Ni–P alloy coatings with tunable phosphorus contents are electrodeposited onto a platinum electrode at room temperature using choline chloride, ethylene glycol (1:2 molar ratio) as a deep eutectic solvent, and $NiCl_2·6H_2O$, $NaH_2PO_2·H_2O$ as nickel and phosphorus sources, respectively. Cyclic voltammetry shows that the presence of $H_2PO_2^-$ in the ionic liquid in the case of Ni plating promotes the initiation of Ni–P nucleation [79]. The structure of Ni–P deposits is converted from a crystalline to an amorphous structure with the increase in phosphorus content in the coating.

There is no unambiguous opinion regarding the use of DES for the electrodeposition of metals and alloys [15,71,72]. At present, the comparison of industrial production of functional materials in DES and in aqueous solutions is not in favor of the former. From a practical point of view, DESs application for the deposition of coatings based on most of the metals (iron, copper, brass, nickel, and their alloys) used in the industry to produce functional materials is yet not realistic [15,72]. However, in the future, the use of DES as an electrolytic medium may open up opportunities for obtaining alloys of new compositions with unique physicochemical properties [71].

The main advantages and disadvantages of water and anhydrous electrolytes for metal electrodeposition can be summarized on the basis of the analysis of literature data (Table 3).

Table 3. Main advantages and disadvantages of classic (water) and modern electrolytes for electrodeposition.

Electrolyte	Advantages	Disadvantages	Ref.
Classic water solutions	low temperature, low cost, fast, good adhesion of deposit, easy to control thickness of the deposit, wide range of metal salt concentration	pH dependent, low current efficiency and porosity due to hydrogen evolution. High concentration of toxic component, the anion formation in the solvent decomposition process can lead to the occurrence of a precipitation with the metal ion	[26,70]
ILs	pH independent, deposition metals that are not accessible with conventional aqueous solution, good solvent for organic and inorganic substances	High cost, toxic, different impurities (water et al.)	[3,4,25]
DES	Deposition metals that are not accessible with conventional aqueous solution, the anion formation in the solvent decomposition process can not lead to the occurrence of a precipitation with the metal ion, low cost, non toxic, easy preparation, high purity	There are few systematic studies	[15,20,23]

4. Conclusions

- The relevance of electrodeposited alloys with a wide range of useful properties, such as corrosive, electrocatalytic, and magnetic properties, is shown. Due to these properties, the scope of applications for these alloys is very wide. The ability to deposit alloys on parts of a complex shape increases the wear and corrosion resistance of the latter.
- It is very important to study the correlation between the concentration of metal ions in the electrolytic bath and the metal content in the reduced form to describe the technique of manufacturing the coatings.
- Also promising, in our opinion, will be the establishment of a correlation between the magnetic and catalytic properties of electrodeposited alloys based on iron triad metals. Another promising direction of research on the properties of electrodeposited nickel- and cobalt-based alloys with modifying additives is the use of these materials in medicine, for example, as a replacement for platinum cardiac stents [61–64].
- The potentiostatic, ultrasonic, and plasma generation techniques in liquid are promising approaches in the electrodeposition process. Solution plasma sputtering is a simple

- and facile technique for the preparation of highly dispersed nanomaterials (metals and alloys) on a variety of carriers, such as metal oxides and carbon materials.
- The electrochemical deposition of alloys based on Fe-triad with aqueous (classical) solvents due to their advantages (Table 3) is today the most commonly used technology for industry. Due to their specific physicochemical and electrochemical properties, ionic liquids and DES are promising for obtaining coatings with new compositions and unique properties that cannot be obtained from classical solvents.

Author Contributions: O.L.—writing and draft preparation of all Parts; and final version preparation; L.F.—writing and draft preparation of Parts 1,2; A.K.—writing and draft preparation of Parts 2; D.K.—writing and draft preparation of Parts 1,3; and final version preparation; L.K.—editing and proof-reading, and final version preparation. All authors have read and agreed to the published version of the manuscript.

Funding: This work was supported by the Ministry of Science and Higher Education of the Russian Federation (grant no. 075-15-2021-591).

Informed Consent Statement: Not applicable.

Data Availability Statement: Not applicable.

Conflicts of Interest: The authors declare no conflict of interest.

Abbreviations

ionic liquids (ILs), deep eutectic solvents (DES), functionally graded coatings (FGC), hydrogen evolution reaction (HER), oxygen evolution reaction (OER), from differential scanning calorimetry (DSC), X-ray emission spectroscopy (vtc-XES), high-entropy alloy (HEA), direct current (DC), pulse current (PC), choline chloride (ChCl), ethylene glycol (EG).

References

1. Zangari, G. Electrodeposition of Alloys and Compounds in the Era of Microelectronics and Energy Conversion Technology. *Coatings* **2015**, *5*, 195–218. [CrossRef]
2. Ma, L.; Xiaoli, X.; Nie, Z.; Dong, T.; Mao, Y. Electrodeposition and Characterization of Co-W Alloy from Regenerated Tungsten Salt. *Int. J. Electrochem. Sci.* **2017**, *12*, 1034–1051. [CrossRef]
3. Li, Y.; Cai, X.; Zhang, G.; Xu, C.; Guo, W.; An, M. Optimization of Electrodeposition Nanocrytalline Ni-Fe Alloy Coatings for the Replacement of Ni Coatings. *J. Alloys Compd.* **2022**, *903*, 163591. [CrossRef]
4. Li, A.; Zhu, Z.; Xue, Z.; Liu, Y. Periodic Ultrasound-Assisted Electrodeposition of Fe–Ni Alloy Foil. *Mater. Res. Bull.* **2022**, *150*, 111778. [CrossRef]
5. Safonov, V.A.; Habazaki, H.; Glatzel, P.; Fishgoit, L.A.; Drozhzhin, O.A.; Lafuerza, S.; Safonova, O.V. Application of Valence-to-Core X-Ray Emission Spectroscopy for Identification and Estimation of Amount of Carbon Covalently Bonded to Chromium in Amorphous Cr-C Coatings Prepared by Magnetron Sputtering. *Appl. Surf. Sci.* **2018**, *427*, 566–572. [CrossRef]
6. Knyazev, A.V.; Fishgoit, L.A.; Chernavskii, P.A.; Safonov, V.A.; Filippova, S.E. Magnetic Properties of Electrodeposited Ni–P Alloys with Varying Phosphorus Content. *Russ. J. Phys. Chem.* **2017**, *91*, 260–263. [CrossRef]
7. Knyazev, A.V.; Fishgoit, L.A.; Chernavskii, P.A.; Safonov, V.A.; Filippova, S.E. Magnetic Properties of Electrodeposited Amorphous Nickel–Phosphorus Alloys. *Russ. J. Electrochem.* **2017**, *53*, 270–274. [CrossRef]
8. Liu, S.; Shohji, I.; Kobayashi, T.; Hirohashi, J.; Wake, T.; Yamamoto, H.; Kamakoshi, Y. Mechanistic Study of Ni–Cr–P Alloy Electrodeposition and Characterization of Deposits. *J. Electroanal. Chem.* **2021**, *897*, 115582. [CrossRef]
9. Jiang, W.; Li, H.; Lao, Y.; Li, X.; Fang, M.; Chen, Y. Synthesis and Characterization of Amorphous NiCoP Alloy Films by Magnetic Assisted Jet Electrodeposition. *J. Alloys Compd.* **2022**, *910*, 164848. [CrossRef]
10. Safonov, V.A.; Safonova, O.V.; Fishgoit, L.A.; Kvashnina, K.; Glatzel, P. Chemical State of Phosphorus in Amorphous Ni–Fe–P Electroplates. *Surf. Coat. Technol.* **2015**, *275*, 239–244. [CrossRef]
11. Safonov, V.A.; Fishgoit, L.A.; Safonova, O.V.; Glatzel, P. On the Presence of Covalently Bound Phosphorus in Amorphous Ni–Co–P and Fe–Co–P Electroplates. *Mater. Chem. Phys.* **2021**, *272*, 124987. [CrossRef]
12. Ranjan, P.; Kumar, R.; Walia, R.S. Functionally Graded Material Coatings (FGMC)—A Review. *J. Phys. Conf. Ser.* **2021**, *2007*, 012068. [CrossRef]
13. Moiseev, I.I.; Loktev, A.S.; Shlyakhtin, O.A.; Mazo, G.N.; Dedov, A.G. New Approaches to the Design of Nickel, Cobalt, and Nickel–Cobalt Catalysts for Partial Oxidation and Dry Reforming of Methane to Synthesis Gas. *Pet. Chem.* **2019**, *59*, S1–S20. [CrossRef]

14. Lebedeva, O.; Kultin, D.; Kalmykov, K.; Snytko, V.; Kuznetsova, I.; Orekhov, A.; Zakharov, A.; Kustov, L. Nanorolls Decorated with Nanotubes as a Novel Type of Nanostructures: Fast Anodic Oxidation of Amorphous Fe–Cr–B Alloy in Hydrophobic Ionic Liquid. *ACS Appl. Mater. Interfaces* **2021**, *13*, 2025–2032. [CrossRef] [PubMed]
15. Bernasconi, R.; Panzeri, G.; Accogli, A.; Liberale, F.; Nobili, L.; Magagnin, L. Electrodeposition from Deep Eutectic Solvents. In *Progress and Developments in Ionic Liquids*; Handy, S., Ed.; InTechOpen: London, UK, 2017; ISBN 978-953-51-2901-1.
16. Yamasaki, T. High-Strength Nanocrystalline Ni-W Alloys Produced by Electrodeposition and Their Embrittlement Behaviors during Grain Growth. *Scr. Mater.* **2001**, *44*, 1497–1502. [CrossRef]
17. Quiroga Argañaraz, M.P.; Ribotta, S.B.; Folquer, M.E.; Benítez, G.; Rubert, A.; Gassa, L.M.; Vela, M.E.; Salvarezza, R.C. The Electrochemistry of Nanostructured Ni–W Alloys. *J. Solid State Electrochem.* **2013**, *17*, 307–313. [CrossRef]
18. Wang, Z.; Wu, T.; Geng, X.; Ru, J.; Hua, Y.; Bu, J.; Xue, Y.; Wang, D. The Role of Electrolyte Ratio in Electrodeposition of Nanoscale Fe Cr Alloy from Choline Chloride-Ethylene Glycol Ionic Liquid: A Suitable Layer for Corrosion Resistance. *J. Mol. Liq.* **2022**, *346*, 117059. [CrossRef]
19. Omar, I.M.A.; Al-Fakih, A.M.; Aziz, M.; Emran, K.M. Part II: Impact of Ionic Liquids as Anticorrosives and Additives on Ni-Co Alloy Electrodeposition: Experimental and DFT Study. *Arab. J. Chem.* **2021**, *14*, 102909. [CrossRef]
20. Maizi, R. Electrodeposition of Ni, Fe and Ni-Fe Alloys in Two Ionic Liquids: (Tri (n-Butyl) [2-Methoxy-2-Oxoethyl] Ammonium Bis (Trifluoromethylsulfonyl) [BuGBOEt] [Tf2N] and (1-Butyl-1- Methylpyrrolidinium Bis Trifluoromethylsulfonyl) Imide ([P1,4] [Tf2N]). *Int. J. Electrochem. Sci.* **2016**, *11*, 7111–7124. [CrossRef]
21. Wallace, A.G.; Symes, M.D. Water-Splitting Electrocatalysts Synthesized Using Ionic Liquids. *Trends Chem.* **2019**, *1*, 247–258. [CrossRef]
22. Zhou, J.; Meng, X.; Ouyang, P.; Zhang, R.; Liu, H.; Xu, C.; Liu, Z. Electrochemical Behavior and Electrodeposition of Fe-Co-Ni Thin Films in Choline Chloride/Urea Deep Eutectic Solvent. *J. Electroanal. Chem.* **2022**, *919*, 116516. [CrossRef]
23. Lebedeva, O.; Kultin, D.; Kustov, L. Electrochemical Synthesis of Unique Nanomaterials in Ionic Liquids. *Nanomaterials* **2021**, *11*, 3270. [CrossRef] [PubMed]
24. Lebedeva, O.; Kultin, D.; Zakharov, A.; Kustov, L. Advances in Application of Ionic Liquids: Fabrication of Surface Nanoscale Oxide Structures by Anodization of Metals and Alloys. *Surf. Interfaces* **2022**, *34*, 102345. [CrossRef]
25. Liu, F.; Deng, Y.; Han, X.; Hu, W.; Zhong, C. Electrodeposition of Metals and Alloys from Ionic Liquids. *J. Alloys Compd.* **2016**, *654*, 163–170. [CrossRef]
26. Costa, J.G. dos R. da; Costa, J.M.; Almeida Neto, A.F. de Progress on Electrodeposition of Metals and Alloys Using Ionic Liquids as Electrolytes. *Metals* **2022**, *12*, 2095. [CrossRef]
27. Smith, E.L.; Abbott, A.P.; Ryder, K.S. Deep Eutectic Solvents (DESs) and Their Applications. *Chem. Rev.* **2014**, *114*, 11060–11082. [CrossRef]
28. Brenner, A. *Electrodeposition of Alloys*; Principles and Practice; Academic Press: New York, NY, USA; London, UK, 1963; Volume 1.
29. Landolt, D. Fundamental aspects of alloy plating. *Plat. Surf. Finish.* **2001**, *88*, 70–79.
30. Cao, X.; Wang, H.; Liu, T.; Shi, Y.; Xue, X. Electrodeposition of Bi from Choline Chloride-Malonic Acid Deep Eutectic Solvent. *Materials* **2023**, *16*, 415. [CrossRef]
31. Lv, Q.; Yao, B.; Zhang, W.; She, L.; Ren, W.; Hou, L.; Fautrelle, Y.; Lu, X.; Yu, X.; Li, X. Controlled Direct Electrodeposition of Crystalline NiFe/Amorphous NiFe-(Oxy)Hydroxide on NiMo Alloy as a Highly Efficient Bifunctional Electrocatalyst for Overall Water Splitting. *Chem. Eng. J.* **2022**, *446*, 137420. [CrossRef]
32. Saeki, R.; Yakita, T.; Ohgai, T. Magnetization and Microhardness of Iron–Chromium Alloy Films Electrodeposited from an Aqueous Solution Containing N, N-Dimethylformamide. *J. Mater. Res. Technol.* **2022**, *18*, 2735–2744. [CrossRef]
33. Torabinejad, V.; Aliofkhazraei, M.; Assareh, S.; Allahyarzadeh, M.H.; Rouhaghdam, A.S. Electrodeposition of Ni-Fe Alloys, Composites, and Nano Coatings—A Review. *J. Alloys Compd.* **2017**, *691*, 841–859. [CrossRef]
34. Shetty, A.R.; Hegde, A.C. Effect of Magnetic Field on Corrosion Performance of Ni-Co Alloy Coatings. *J. Bio- Tribo-Corros.* **2023**, *9*, 16. [CrossRef]
35. Haché, M.J.R.; Tam, J.; Erb, U.; Zou, Y. Electrodeposited Nanocrystalline Medium-Entropy Alloys—An Effective Strategy of Producing Stronger and More Stable Nanomaterials. *J. Alloys Compd.* **2022**, *899*, 163233. [CrossRef]
36. Barati Darband, G.; Aliofkhazraei, M.; Rouhaghdam, A.S. Facile Electrodeposition of Ternary Ni-Fe-Co Alloy Nanostructure as a Binder Free, Cost-Effective and Durable Electrocatalyst for High-Performance Overall Water Splitting. *J. Colloid Interface Sci.* **2019**, *547*, 407–420. [CrossRef] [PubMed]
37. Bachvarov, V.; Lefterova, E.; Rashkov, R. Electrodeposited NiFeCo and NiFeCoP Alloy Cathodes for Hydrogen Evolution Reaction in Alkaline Medium. *Int. J. Hydrogen Energy* **2016**, *41*, 12762–12771. [CrossRef]
38. Kothanam, N.; Harachai, K.; Qin, J.; Boonyongmaneerat, Y.; Triroj, N.; Jaroenapibal, P. Hardness and Tribological Properties of Electrodeposited Ni–P Multilayer Coatings Fabricated through a Stirring Time-Controlled Technique. *J. Mater. Res. Technol.* **2022**, *19*, 1884–1896. [CrossRef]
39. Aliofkhazraei, M.; Walsh, F.C.; Zangari, G.; Köçkar, H.; Alper, M.; Rizal, C.; Magagnin, L.; Protsenko, V.; Arunachalam, R.; Rezvanian, A.; et al. Development of Electrodeposited Multilayer Coatings: A Review of Fabrication, Microstructure, Properties and Applications. *Appl. Surf. Sci. Adv.* **2021**, *6*, 100141. [CrossRef]
40. Ma, C.; Wang, S.; Walsh, F.C. The Electrodeposition of Nanocrystalline Cobalt–Nickel–Phosphorus Alloy Coatings: A Review. *Trans. Inst. Met. Finish.* **2015**, *93*, 275–280. [CrossRef]

41. Filgueira de Almeida, A.; Venceslau de Souto, J.I.; Lima dos Santos, M.; Costa de Santana, R.A.; Alves, J.J.N.; Nascimento Campos, A.R.; Prasad, S. Establishing Relationships between Bath Composition and the Properties of Amorphous Ni–Mo Alloys Obtained by Electrodeposition. *J. Alloys Compd.* **2021**, *888*, 161595. [CrossRef]
42. Benavente Llorente, V.; Diaz, L.A.; Lacconi, G.I.; Abuin, G.C.; Franceschini, E.A. Effect of Duty Cycle on NiMo Alloys Prepared by Pulsed Electrodeposition for Hydrogen Evolution Reaction. *J. Alloys Compd.* **2022**, *897*, 163161. [CrossRef]
43. Shojaei, Z.; Khayati, G.R.; Darezereshki, E. Review of Electrodeposition Methods for the Preparation of High-Entropy Alloys. *Int. J. Miner. Metall. Mater.* **2022**, *29*, 1683–1696. [CrossRef]
44. Yue, Z.; Muig, M.; Xiao-Ming, X.; Ze-Lin, L.; Shi-Xun, L.; Sbao-Min, Z. Kinetic Model of Induced Codeposition of Ni-Mo Alloys. *Chin. J. Chem.* **2010**, *18*, 29–34. [CrossRef]
45. Jović, B.M.; Jović, V.D.; Maksimović, V.M.; Pavlović, M.G. Characterization of Electrodeposited Powders of the System Ni–Mo–O. *Electrochim. Acta* **2008**, *53*, 4796–4804. [CrossRef]
46. Podlaha, E.J.; Landolt, D. Induced Codeposition: II. A Mathematical Model Describing the Electrodeposition of Ni-Mo Alloys. *J. Electrochem. Soc.* **1996**, *143*, 893–899. [CrossRef]
47. Manazoğlu, M.; Hapçı, G.; Orhan, G. Electrochemical Deposition and Characterization of Ni-Mo Alloys as Cathode for Alkaline Water Electrolysis. *J. Mater. Eng. Perform.* **2016**, *25*, 130–137. [CrossRef]
48. Tsyntsaru, N.; Cesiulis, H.; Donten, M.; Sort, J.; Pellicer, E.; Podlaha-Murphy, E.J. Modern Trends in Tungsten Alloys Electrodeposition with Iron Group Metals. *Surf. Eng. Appl. Electrochem.* **2012**, *48*, 491–520. [CrossRef]
49. Fishgoit, L.A.; Fedorayev, I.I.; Knyazev, A.V.; Kasyanov, F.V.; Perkovskii, E.A. Electrodeposition of Nanocoatings Involving Iron Triad Metals. *Gal'vanotekh. Obrab. Poverkhn.* **2022**, *30*, 13–28. [CrossRef]
50. Zhu, Z.; Meng, H.; Ren, P. CoNiWReP High Entropy Alloy Coatings Prepared by Pulse Current Electrodeposition from Aqueous Solution. *Colloids Surf. A* **2022**, *648*, 129404. [CrossRef]
51. Zhang, W.; Xia, W.; Li, B.; Li, M.; Hong, M.; Zhang, Z. Influences of Co and Process Parameters on Structure and Corrosion Properties of Nanocrystalline Ni-W-Co Ternary Alloy Film Fabricated by Electrodeposition at Low Current Density. *Surf. Coat. Technol.* **2022**, *439*, 128457. [CrossRef]
52. Crousier, J.; Eyraud, M.; Crousier, J.-P.; Roman, J.-M. Influence of Substrate on the Electrodeposition of Nickel-Molybdenum Alloys. *J. Appl. Electrochem.* **1992**, *22*, 749–755. [CrossRef]
53. Beltowska-Lehman, E.; Chassaing, E. Electrochemical Investigation of the Ni±Cu±Mo Electrodeposition System. *J. Appl. Electrochem.* **1997**, *27*, 568–572. [CrossRef]
54. Allahyarzadeh, M.H.; Aliofkhazraei, M.; Rezvanian, A.R.; Torabinejad, V.; Sabour Rouhaghdam, A.R. Ni-W Electrodeposited Coatings: Characterization, Properties and Applications. *Surf. Coat. Technol.* **2016**, *307*, 978–1010. [CrossRef]
55. Salehikahrizsangi, P.; Raeissi, K.; Karimzadeh, F.; Calabrese, L.; Proverbio, E. Highly Hydrophobic Ni-W Electrodeposited Film with Hierarchical Structure. *Surf. Coat. Technol.* **2018**, *344*, 626–635. [CrossRef]
56. Lee, S.; Choi, M.; Park, S.; Jung, H.; Yoo, B. Mechanical Properties of Electrodeposited Ni-W ThinFilms with Alternate W-Rich and W-Poor Multilayers. *Electrochim. Acta* **2015**, *153*, 225–231. [CrossRef]
57. Slavcheva, E.; Mokwa, W.; Schnakenberg, U. Electrodeposition and Properties of NiW Films for MEMS Application. *Electrochim. Acta* **2005**, *50*, 5573–5580. [CrossRef]
58. Zhang, Z.; Xu, Z.; Liao, Z.; Chen, C.; Wei, G. A Novel Synthesis Method for Functionally Graded Alloy Coatings by Induced Electrodeposition. *Mater. Lett.* **2022**, *312*, 131681. [CrossRef]
59. Zhang, Z.; Dai, L.; Yin, Y.; Xu, Z.; Lv, Y.; Liao, Z.; Wei, G.; Zhong, F.; Yuan, M. Electrodeposition and Wear Behavior of NiCoW Ternary Alloy Coatings Reinforced by Al_2O_3 Nanoparticles: Influence of Current Density and Electrolyte Composition. *Surf. Coat. Technol.* **2022**, *431*, 128030. [CrossRef]
60. Cesiulis, H.; Tsyntsaru, N.; Budreika, A.; Skridaila, N. Electrodeposition of CoMo and CoMoP Alloys from the Weakly Acidic Solutions. *Surf. Eng. Appl. Electrochem.* **2010**, *46*, 406–415. [CrossRef]
61. Huang, H.; Hu, X.; Zhang, J.; Su, N.; Cheng, J. Facile Fabrication of Platinum-Cobalt Alloy Nanoparticles with Enhanced Electrocatalytic Activity for a Methanol Oxidation Reaction. *Sci. Rep.* **2017**, *7*, 45555. [CrossRef]
62. Saito, G.; Akiyama, T. Nanomaterial Synthesis Using Plasma Generation in Liquid. *J. Nanomater.* **2015**, *16*, 299. [CrossRef]
63. Endres, F.; Abbott, A.; MacFarlane, D. (Eds.) *Electrodeposition from Ionic Liquids*, 2nd ed.; Wiley-VCH Verlag GmbH & Co. KGaA: Weinheim, Germany, 2017; ISBN 978-3-527-68270-6.
64. Greer, A.J.; Jacquemin, J.; Hardacre, C. Industrial Applications of Ionic Liquids. *Molecules* **2020**, *25*, 5207. [CrossRef] [PubMed]
65. Ispas, A.; Bund, A. Electrodeposition in Ionic Liquids. *Electrochem. Soc. Interface* **2014**, *23*, 47–51. [CrossRef]
66. Omar, I.M.A.; Aziz, M.; Emran, K.M. Part I: Ni-Co Alloy Foils Electrodeposited Using Ionic Liquids. *Arab. J. Chem.* **2020**, *13*, 7707–7719. [CrossRef]
67. Mohanty, U.S.; Tripathy, B.C.; Singh, P.; Keshavarz, A.; Iglauer, S. Roles of Organic and Inorganic Additives on the Surface Quality, Morphology, and Polarization Behavior during Nickel Electrodeposition from Various Baths: A Review. *J. Appl. Electrochem.* **2019**, *49*, 847–870. [CrossRef]
68. Danilov, F.I.; Bogdanov, D.A.; Smyrnova, O.V.; Korniy, S.A.; Protsenko, V.S. Electrodeposition of Ni–Fe Alloy from a Choline Chloride-Containing Ionic Liquid. *J. Solid State Electrochem.* **2022**, *26*, 939–957. [CrossRef]
69. Tułodziecki, M.; Tarascon, J.-M.; Taberna, P.L.; Guéry, C. Importance of the Double Layer Structure in the Electrochemical Deposition of Co from Soluble Co2+—Based Precursors in Ionic Liquid Media. *Electrochim. Acta* **2014**, *134*, 55–66. [CrossRef]

70. Brenner, A. *Electrodeposition of Alloys. 2: Practical and Specific Information*; Academic Press: New York, NY, USA, 1963; ISBN 978-1-4832-0967-8.
71. Abbott, A.P. Deep Eutectic Solvents and Their Application in Electrochemistry. *Curr. Opin. Green Sustain. Chem.* **2022**, *36*, 100649. [CrossRef]
72. Smith, E.L. Deep Eutectic Solvents (DESs) and the Metal Finishing Industry: Where Are They Now? *Trans. Inst. Met. Finish.* **2013**, *91*, 241–248. [CrossRef]
73. Gómez, E.; Fons, A.; Cestaro, R.; Serrà, A. Electrodeposition of CoNi Alloys in a Biocompatible DES and Its Suitability for Activating the Formation of Sulfate Radicals. *Electrochim. Acta* **2022**, *435*, 141428. [CrossRef]
74. Landa-Castro, M.; Sebastián, P.; Giannotti, M.I.; Serrà, A.; Gómez, E. Electrodeposition of Nanostructured Cobalt Films from a Deep Eutectic Solvent: Influence of the Substrate and Deposition Potential Range. *Electrochim. Acta* **2020**, *359*, 136928. [CrossRef]
75. Oliveira, F.G.S.; Santos, L.P.M.; da Silva, R.B.; Correa, M.A.; Bohn, F.; Correia, A.N.; Vieira, L.; Vasconcelos, I.F.; de Lima-Neto, P. FexNi(1-x) Coatings Electrodeposited from Choline Chloride-Urea Mixture: Magnetic and Electrocatalytic Properties for Water Electrolysis. *Mater. Chem. Phys.* **2022**, *279*, 125398. [CrossRef]
76. Higashino, S.; Abbott, A.P.; Miyake, M.; Hirato, T. Iron(III) Chloride and Acetamide Eutectic for the Electrodeposition of Iron and Iron Based Alloys. *Electrochim. Acta* **2020**, *351*, 136414. [CrossRef]
77. Niciejewska, A.; Ajmal, A.; Pawlyta, M.; Marczewski, M.; Winiarski, J. Electrodeposition of Ni–Mo Alloy Coatings from Choline Chloride and Propylene Glycol Deep Eutectic Solvent Plating Bath. *Sci. Rep.* **2022**, *12*, 18531. [CrossRef] [PubMed]
78. Ysea, N.B.; Benavente Llorente, V.; Loiácono, A.; Lagucik Marquez, L.; Diaz, L.; Lacconi, G.I.; Franceschini, E.A. Critical Insights from Alloys and Composites of Ni-Based Electrocatalysts for HER on NaCl Electrolyte. *J. Alloys Compd.* **2022**, *915*, 165352. [CrossRef]
79. You, Y.; Gu, C.; Wang, X.; Tu, J. Electrochemical Synthesis and Characterization of Ni–P Alloy Coatings from Eutectic–Based Ionic Liquid. *J. Electrochem. Soc.* **2012**, *159*, D642–D648. [CrossRef]

Disclaimer/Publisher's Note: The statements, opinions and data contained in all publications are solely those of the individual author(s) and contributor(s) and not of MDPI and/or the editor(s). MDPI and/or the editor(s) disclaim responsibility for any injury to people or property resulting from any ideas, methods, instructions or products referred to in the content.

Article

Influence of Pretreatment Processes on Adhesion of Ni/Cu/Ni Multilayer on Polyetherimide Resin Reinforced with Glass Fibers

Xiaodong Xu [1], Dingkai Xie [1], Jiaqi Huang [1], Kunming Liu [2], Guang He [1], Yi Zhang [1], Peng Jiang [1], Lixin Tang [3] and Wangping Wu [1],*

1. Electrochemistry and Corrosion Laboratory, School of Mechanical Engineering and Rail Transit, Changzhou University, Changzhou 213164, China
2. Jiangsu Kexiang Anticorrosive Materials Co., Ltd., Changzhou 213100, China
3. Zhenjiang Arf Special Coating Technology Co., Ltd., Zhenjiang 212006, China
* Correspondence: wwp3.14@163.com or wuwping@cczu.edu.cn

Abstract: The metallization of polyetherimide (PEI) is widely considered to enhance its surface properties and enhance its application in engineering fields; however, adhesion is a key factor in determining the reliability of PEI metallization. A Ni/Cu/Ni multilayer coating was successfully manufactured on a batch of PEI resin reinforced with glass fibers by a two-step metallization process, including sandblasting and activation/acceleration. The microstructure and morphology of the top-surface and cross-section of the coatings were observed by scanning electron microscopy. The chemical state and composition of the deposits were characterized by both X-ray photoelectron and energy-dispersive spectroscopy. The adhesion state was qualitatively evaluated by cross-cut tests with 3M tape. The surface roughness of the substrate significantly increased after the sandblasting process, which could improve the adhesion between the multilayer coating and the PEI substrate. After the standard activation process, the acceleration made an effect on the deposition of the initial Ni layer for electroless plating. The influence of acceleration on the appearance quality of metallization on the PEI substrate was studied and, at the same time, the mechanism of acceleration was investigated and addressed.

Keywords: pretreatment; polyetherimide; electroless plating; metallization; adhesion

1. Introduction

Currently, the surface metallization of polymeric substrates, with or without fiber reinforcement, is a study area of growing interest because of the need to enhance their surface properties. In fact, the polymer-based materials are increasingly used in several sectors of engineering, such as automotive, aerospace, and construction, replacing metals for different applications [1]. Polyetherimide (PEI)-based materials are amorphous, thermoplastic polymers, which are chemically stable, biocompatible, and exhibit the mechanical, thermal, and dielectric properties required for next-generation applications [2]. PEI-based materials have some advantages, such as high heat resistance, stiffness, impact strength, high mechanical strength and flame resistance, and broad chemical resistance [3,4]. PEI could offer advantages such as lightness, a high strength to weight ratio, and flexibility in designing shapes and forms. On the other hand, it could also be useful to improve some of its properties, such as its electrical conductivity, electromagnetic shielding capabilities, thermal conductivity, flame resistance, and erosion and radiation protection, in order to further widen the fields of application for PEI-based materials. In this regard, the surface metallization of the PEI substrate is an effective technique to enhance the above-mentioned surface properties to expand their engineering applications [5].

Some approaches could be used to prepare a metallic layer on the polymers, such as physical (PVD) [6,7] and chemical vapor deposition (CVD) [8,9], the use of thermal [10]

and cold spray [1,11], using electroless plating [12–14], and electroplating a conductive layer [15]. PVD, CVD, and spray techniques incur high equipment and processing costs, and are limited in terms of productivity and workpiece sizes and shapes. Electroless plating is an alternative way to produce the metallic layer on PEI composite substrates when plating without applied current and vacuum conditions or using a low bath temperature, and has applicability to complicated-shaped substrates [16–19]. Electroless nickel plating on polyethylene terephthalate surfaces for triboelectric nanogenerators facilitates the improvement of device wearability and comfort [20]. Marques-Hueso et al. [21] reported that a rapid photosynthesis method was used to produce the silver nanoparticles on PEI substrates. Marline et al. [22] studied a direct metallization of Ni coating on PEI substrates by spin-coating or a dipping deposition of an ultra-thin film of an organic nickel salt in an alcoholic solution. Jones et al. [23] investigated a direct metallization of PEI substrates by activation with metal ions as seed layers, which could be reduced and formed by chemical or optical reduction. Alodan [24] reported a metallizing PEI resin reinforced with glass fibers, where the surface of the PEI substrate was modified by etching the resin matrix and the glass fibers using two different solutions. During the resin etching step, glass fibers were exposed and etched at the surface. Glass fiber etching was found to be essential to promote adhesion between the metallic film and the PEI surface. After electroplating, the adhesion force was 230 kg/cm^2 (3000 psi). Esfahani et al. [2] presented a new digital fabrication strategy that combined the 3D printing of high-performance polymers (polyetherimide) with a light-based selective metallization of copper traces through the chemical modification of the polymer surface and the computer-controlled assembly of functional devices and structures. Zhang et al. [25] reported that a new surface modification method was developed for the electroless deposition of robust metal (copper, nickel, silver, etc.) layers on a poly(dimethylsiloxane) substrate with strong adhesion.

Adhesion strength is an important parameter in determining the reliability of the metallization of the PEI substrates [26]. In fact, it is hard to obtain a strong adhesion between the layer and the PEI substrate due to high interfacial energy. Some pretreatment processes could be helpful to improve the adhesion strength, including laser [27–29], plasma [30,31], and mechanical roughening [32,33], as well as chemical etching [34,35]. These pretreatment processes increase the surface roughness, and wettability or chemical bonding of the substrates, resulting in a strong adhesion between the metallic layer and the substrate. For conventional electroless plating, adsorbed palladium (Pd) catalysts are widely used to activate the surface for the following electroless metallization. The sensitization/activation and acceleration could seed a catalyst on the surface of the PEI substrates for electroless plating, which could also influence the adhesion of the metallic layer. Pretreatment of the PEI substrate with the PdCl/SnCl sensitizer solution leads to the adsorption of catalytically inactive Sn and Pd compounds or complexes [36]. Subsequently, the excess tin oxides, hydroxides, and salts were removed by an accelerator solution, and then the active catalyst species consisted of left behind metallic Pd and PdSn particles approximately 1–10 nm in size, which become the deposition center of Ni atoms for electroless plating. However, there are few reports on the influence of pretreatment processes, including mechanical sandblasting and acceleration, on the adhesion of the metallization of PEI substrates. Generally, the deposition of electroless plating on PEI substrates required a chemical pretreatment to ensure good adhesion; however, there are limitations due to its low adhesive strength and a long production cycle strongly linked to environmental hazards/costs. [34,35].

In a recent publication, a composite coating of a Ni/Cu/Ni multilayer was produced on the surface of PEI reinforced with glass fiber composite substrates. Ni and Cu layers were obtained by electroless plating and electrodeposition, respectively. The influences of the Cu interlayer thickness and heat treatment on the adhesion state of the multilayer coatings on the substrate were studied [32]. Film thickness has an important effect on adhesion. The adhesion solution of the Ni/7.54 μm Cu/Ni and Ni/58.6 μm Cu/Ni multilayers on PEI composites could be classified as grade 5B and 3B, respectively. Furthermore, the adhesion force of the Ni/58.6 μm Cu/Ni multilayer on the substrate can be improved by heat treatment at 200 °C for 4 h. However, the pretreatment processes for metallization on PEI substrates play a key factor on the adhesion and appearance quality of the deposits. In this study, a coating of Ni/Cu/electroless Ni multilayer was deposited on PEI reinforced with the glass fibers. Therefore, it is necessary to study the effect of pretreatment processes, such as roughening and acceleration processes, on the adhesion of the Ni/Cu/Ni multilayer on PEI composite substrates. Tape peel tests were used to determine the adhesion state of Ni/Cu/Ni/PEI composites with pretreatment processes.

2. Experimental Section

2.1. Pretreatment and Preparation

We manufactured the PEI composite substrates with short glass fibers by injection molding, which could be used as electronic products in the communication field. The two-dimensional drawings of the PEI composite substrate are displayed in Figure 1.

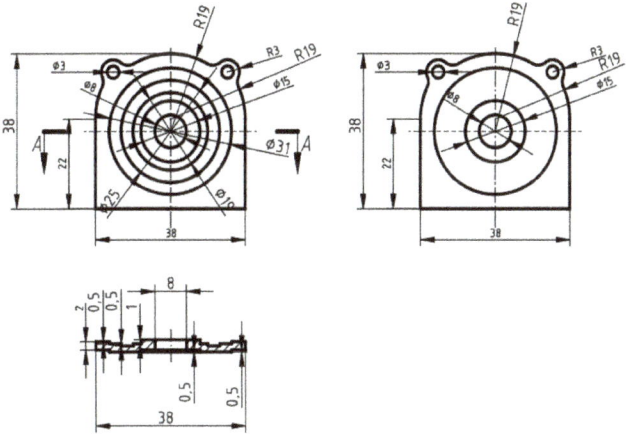

Figure 1. Two-dimensional drawings of PEI composite substrate.

Figure 2 shows the process flowchart for surface metallization on PEI substrates. The preparation of metallization of the PEI substrate was performed in the subsequent steps: (1) heat treatment, (2) machine coarsening, (3) ultrasonic degrease, (4) chemical etching, (5) surface neutralizing, (6) sensitization and activation, (7) acceleration, (8) alkaline nickel electroless plating, (9) weak corrosion, (10) copper electroplating, and (11) acid electroless nickel plating. After each step, we washed and rinsed the substrates in deionized water. Table 1 displays each process involved in metallization on PEI substrates. The details for each step are as follows:

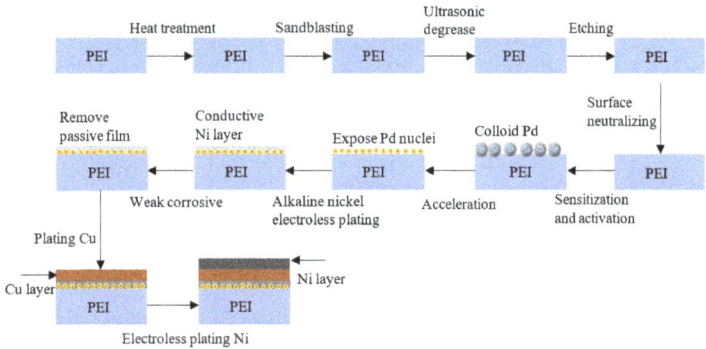

Figure 2. Process flowchart for metallization of PEI composite substrates.

Table 1. Process conditions of surface neutralizing (Modifier A/B/C were purchased from MacDermid Co., Shanghai, China).

Item	Condition
Modifier A	7–10%
Modifier B	5–6%
Modifier C	1.5–3%
Temperature (°C)	50–60
Time (min)	12–18
pH	9–10

Note: Mixed solution is of deionized water.

Heat treatment: We heat treated PEI substrates at 50~100 °C for 30 min, then 200 °C for 4~4.5 h.

Sandblasting: We sandblasted with 5 kg pressure for holding time of 4~5 s to roughen the PEI surface. These sands were composed of white Al_2O_3 grits with sizes ranging from 9.8 to 10.2 μm. The distance between the nozzle and the sample was about 80~90 cm. We immersed the sandblasted samples in deionized water, then took them out and placed them horizontally. The sandblasting resulted in an increase in the surface roughness, improving the adhesion between the layer and the substrate.

Ultrasonic degrease: After sandblasting, we degreased the PEI samples and cleansed in the ultrasonic cleaning machine with alkaline solution 5.0 M NaOH for ~20 min. Then, we rinsed in the deionized water.

Chemical etching: We chemically etched the samples in a mixture of chromium hexavalent, containing chromium trioxide of 200–400 g/L and 10 % sulfuric acid solution, and then we rinsed with water. We set the bath temperature in a range from 55 to 60 °C, and the etching time was 15–25 min.

Surface neutralizing: This step was to remove excess chromate from the surface and to neutralize the substrate because of the residual Cr^{6+} on the sample after chemical etching. We performed this step to avoid affecting the other solution, and to neutralize and adjust the samples according to Table 1. This mixed solution could modify the surface of the PEI substrate after etching, and might supply a positively charged group, which could be helpful to absorb the negatively charged Pd group. For some special polymer composite materials, such as PEI and PI composite polymer materials, this step has to be performed after the etching process to obtain much more Pd species for the next step-activation/sensitization process. However, for the metallization of ABS polymer materials, this step is not performed after the etching process (skip to the activation/sensitization process).

Sensitization and activation: We immersed the PEI substrates in the sensitization and activation solutions containing colloid palladium. The gray black Pd nanoparticles were formed by the redox reaction between palladium chloride ($PdCl_2$) and stannous chloride ($SnCl_2$) in the solution [37]. In this case, the Pd nuclei and nanoparticles surrounded by adsorbed Sn^{2+} species were formed on the surface of PEI substrates. The mechanism of colloidal Pd formation is based on the $SnCl_2 \cdot 2H_2O$ solution, and Sn^{2+} ions in the solution can reduce Pd^{2+} ions to Pd atoms. The equations describing the reaction mechanism are as follows [38]:

$$SnCl_2 + PdCl_2 \rightarrow SnCl_4 + Pd \quad (1)$$

$$Pd^{2+} + Sn^{2+} + 4Cl^- \rightarrow Sn(PdCl_4) \quad (2)$$

Acceleration: Pd nuclei and nanoparticles were surrounded by adsorbed Sn^{2+} species, which could remove colloid to expose the surrounded Pd nuclei on PEI substrates under stirring and air-blowing conditions. It is a strong acid solution with a low pH value. The standard treatment time was 4 min at a temperature of 40 °C as the activated samples were immersed in the acceleration solution.

Alkaline electroless nickel plating: We used an alkaline chemical nickel plating to generate a conductive film at room temperature. When the activated samples were immersed in the solution, the reduction of Ni^{2+} ions occurred, and an initial nickel layer was formed on the PEI substrates. During deposition, after a few minutes, the surface of the substrate slowly became shiny in the solution, inferring that the deposition of the metallic layer occurred. Subsequently, we observed that the reaction intensified on the surface of the samples because lots of small bubbles were released from the bath, and the Ni autocatalysis resulted in a vigorous reaction in the solution. Lastly, the Ni layer was deposited on surface of the PEI substrates. The surface of the samples presented a metallic appearance. The bath chemistry and deposition parameters are listed in Table 2. We controlled the thickness of the layer by the deposition time. The thickness of the initial electroless Ni layer was about 2 µm.

Table 2. Process steps involved in metallization on PEI substrates.

Process Step	Process Condition
Heat treatment	First, T = 50~100 °C for 30 min, then 200 °C for 4~4.5 h.
Sandblasting	5 kg pressure
Degreasing	5 M NaOH at T = 25 °C for ~20 min under ultrasonic condition
Chemical etching	Chromium trioxide of 200–400 g/L and 10% sulfuric acid, 55–60 °C, 15 min
Surface neutralizing	5 M NaOH at T = 25 °C for ~20 min under ultrasonic condition
Sensitization and activation	220–280 mL/L HCl; 3 g/L SnCl$_2$·2 H$_2$O; 2–4 mL/L Activator *; T = 25 °C; t = 10 min
Acceleration	100 mL/L HCl, T = 40 °C, t = 2~10 min
Alkaline electroless Ni	90–110 mL/L 160 A *; 25–35 mL/L 160 B *; 4–6 mL/L NH$_3$·H$_2$O; pH = 8.8–9.2; T = 25 °C and t = 8 min
Weak Corrosion	10% HCl; t = 30 s
Cu electroplating	55–60 g/L CuCO$_3$; 30–40 g/L NaKC$_4$H$_4$O$_6$·4H$_2$O; 250–270 g/L C$_6$H$_8$O$_7$; 10–15 g/L NaHCO$_3$; pH = 8.5–9.5; T = 25 °C; Applied voltage = 2.5 V; t = 15 min
Acid electroless Ni	15–25 g/L Ni(NH$_4$)$_2$(SO$_4$)$_2$; 25–35 g/L NaH$_2$PO$_2$·H$_2$O; 25–35 g/L CH$_3$COONa; 20–30 g/L C$_6$H$_5$Na$_3$O$_7$·2H$_2$O; 1 mg/L CH$_4$N$_2$S; T = 70 °C; pH = 4.5–5; t = 20 min

Note: * Chemicals were purchased from McDermid-Canning Com.; T = temperature, t = applied time.

Weak corrosion: Weak corrosion is one of the key processes before the electrodeposition of copper layer. This is a surface activation process. This step is to remove the passivation film formed on the surface of the initial nickel layer, ensuring that good adhesion between the nickel layer and the copper coating. The weak corrosive solution was composed of an aqueous solution with 10% hydrochloric acid, immersed at room temperature for 30 s.

Cu electroplating: The electrodeposition conditions and bath chemistry for Cu electroplating are displayed in Table 2. We electrodeposited the thick Cu coating on the initial Ni layer in an alkaline solution at room temperature for 25 min (a good electrical conductive layer). The thickness of the Cu interlayer was about 20 µm. We controlled the thickness of the Cu coating by the deposition potential and time. We used two Cu plates as the anode, and used the samples with an initial Ni layer as cathodes. During electroplating, the solution was stirred by continuous bubbles, and the distance of the anode and cathode was about 25 cm.

Acid electroless Ni top layer: We deposited the top-layer Ni, as an anti-corrosive layer, on the surface of Cu/Ni PEI composite by electroless process in acidic solution for 20 min. The deposition conditions and bath chemistry are displayed in Table 2. The thickness of the top-surface Ni layer was about 6 µm.

2.2. Adhesion Force Test

We made sets of 4 × 4 cross-hatched scratches, with an interscratch distance of 1 mm and cross scratches at ~90°, on the deposits after drying. We applied a 3 M tape (610-1PK special test tape produced by 3M Company) over the grid, placing the center of the tape over the grid, and smoothed into place with a finger. After holding on 90 s, we removed the tape by seizing the free end and pulled it off rapidly at 90°. We assessed the adhesion force between the layer and the substrate according to the cross-cut testing standard of

ASTM-D3359-09 (see Table 3). We repeated the cross-cut tests at least twice to evaluate the adhesion force between the substrate and the layer.

Table 3. Classification of test results.

Classification	Description	Appearance of Surface of Crosscut from Which Flaking Has Occurred (Example for Six Parallel Cuts)
5B	The edges of the cuts are completely smooth; none of the squares of the lattice are detached.	
4B	Detachment of small flakes of the coating at the intersections of the cuts. A cross-cut area not greater than 5% is affected.	
3B	The coatings flaked along the edges and/or at the intersections of the cuts. The cross-cut area is >5%; however <15% is affected.	
2B	The coatings flaked along the edges of the cuts partly or wholly on different parts of the squares. The cross-cut area is >15%; however, <35% is affected.	
1B	The coatings flaked along the edges of the cuts in large ribbons and/or some squares detached partly or wholly. The cross-cut area is >35%; however, <65% is affected.	
0B	A cross-cut area of >65% is affected.	

2.3. Characterizations

We evaluated the average surface roughness values of the PEI substrates before and after sandblasting by a surface profilometer (BRUKER ContourGT K0) with a 12.5 μm radius tip.

We recorded contact angles of the PEI substrates before and after sandblasting using a goniometer (FM4000, KRUSS Germany). The liquid we used was distilled water, 5 μL in volume, which we dropped on the surface of the samples in order to measure the contact angle. We determined the final contact angle three times at different places on the surface of the samples, using the well-known Young's Equation (3) for describing the contact angle θ on a solid surface,

$$cos\theta = \frac{(\gamma_{SV} - \gamma_{SL})}{\gamma_{LV}} \tag{3}$$

where γ_{SV}, γ_{SL}, and γ_{LV} are the interfacial free energy per unit area of the solid–vapor, solid–liquid, and liquid–vapor interfaces, respectively. A simple model to characterize the influence of the surface roughness on the wettability of a solid was proposed by Wenzel Equation (4), which can be represented as follows:

$$cos\theta^* = R(\gamma_{SV} - \gamma_{SL})/\gamma_{LV} = Rcos\theta \tag{4}$$

where R is the solid surface roughness, defined as the ratio between the actual surface area of a rough surface to the projected area. The above equation shows the relationship between the contact angle of the flat surface θ and that of the fractal surface θ^*. We determined the surface energy by contact angle measurements using the Young–Dupre Equation (5).

$$\gamma = \frac{\gamma_w}{4}(1 + \cos\theta)^2 \quad (5)$$

where γ is the surface energy of the solid, γ_w = 73 mJ/m^2 is the surface energy of the liquid water, and θ is the measured contact angle.

We observed the microstructure and morphology of the surface and cross-section of the deposits by a scanning electron microscopy (SEM, Zeiss-Supra55). We set the scanning rate to 40 s at an operating voltage of 10 kV. We determined the chemical composition of the deposit by X-ray energy-dispersive spectroscopy (EDS, X-act) detector. We measured the chemical state of ions in the solution using a UV Spectrophotometer (UV-3600 Japan).

We performed X-ray photoelectron spectroscopy (XPS) measurements in ultrahigh vacuum (3.3 × 10^{-8} Pa base pressure) using a 5700 ESCA System (PHI, NY, USA). We irradiated the samples with an Al-Kα monochromatic source (1486.6 eV) and analyzed the outcome electrons by a spherical capacitor analyzer using slit aperture of 0.8 mm. We analyzed the samples at the top-surface. We used the carbon signal for C–1s at 285 eV as an energy reference for the measured peaks. In order to identify the elements on the surface of the film, we performed a low-resolution survey spectrum over a wide energy range (0–1000 eV). We acquired high-resolution spectra with a pass energy of 23.5 eV at increments of 0.1 eV·step^{-1}, to allow precise determination of the position of the peaks and their shape. We performed curve-fitting with Gaussian–Lorentzian function, using the XPSPEAK software. We fixed two fitting parameters—the position of the peak and its full width at half-maximum (FWHM)—within ±0.2 eV.

3. Results and Discussion

Figure 3 displays the 3D images of the surface morphology of the PEI substrate before and after the sandblasting process. The average surface roughness (Ra) values of the surface of the PEI substrate before and after sandblasting were 117 nm and 2125 nm, respectively. The surface roughness increased significantly after the sandblasting process. The surface of the PEI substrate was smooth (Figure 3a); however, the surface of the PEI substrate after sandblasting looked like a "mountain chain" (Figure 3b), indicating the surface became rough, which could contribute to improving the adhesion of the layer and the PEI substrate.

The effect of sandblasting on the wettability of the PEI substrate is observed in Figure 4. The surface of the PEI substrate before sandblasting shows a high wettability at a contact angle of about 84° (Figure 4a). After sandblasting, the contact angle increased to 100° (Figure 4b), indicating a low wettability. The specific surface area significantly increased after sandblasting process. The calculated surface energies of the PEI substrates before and after sandblasting were equal to 34.3 mJ/m^2 and 60 mJ/m^2, respectively. This indicated that the specific surface area significantly increased after the sandblasting process. The Wenzel and Cassie models contributed to the effect of surface roughness on wettability. The Wenzel model shows that a rough material surface has a higher surface area than a smooth one, which increases its hydrophobicity. The Cassie model shows that air trapped on the rough surface enhances the hydrophobicity because the drop is partially sitting on air. According to Equations (4) and (5), the ratio R is greater than unity, thus the wettability increases ($\theta^* < \theta$) for a hydrophilic situation, and decreases ($\theta^* > \theta$) for a hydrophobic one. Therefore, this result agrees with the Wenzel equation. However, the surface wettability of the PEI substrate after sandblasting pretreatment can be improved to some degree after the chemical etching process. A good surface wettability of the PEI substrate can enhance the absorption of the activated Pd species in an activation solution. Therefore, more activated Pd particles were fixed and attached at the top-wetted surface, resulting in the increase of surface reactive sites. It can be inferred that the adhesion of the Ni layer was enhanced by the increase of the surface reactive sites.

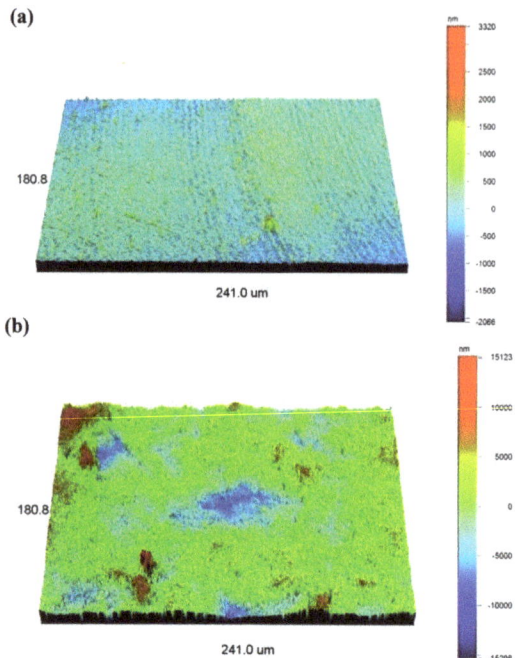

Figure 3. 3D images of surface roughness for PEI composite before (**a**) and after (**b**) sandblasting process.

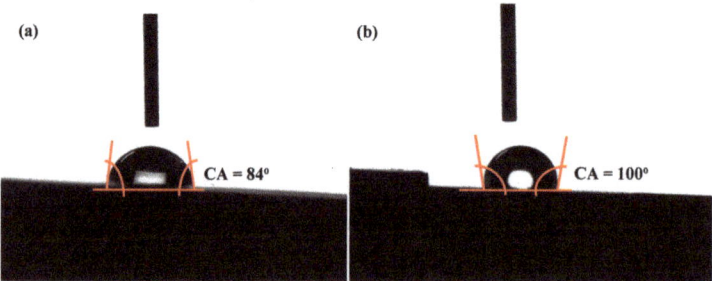

Figure 4. Water droplet contact angles on surface of PEI composite before (**a**) and after (**b**) sandblasting process.

The adhesion between the Ni/Cu/Ni multilayer and the polymer substrate was influenced by the sandblasting process. Figure 5 shows the digital images of the Ni/Cu/Ni deposits on PEI substrates with and without sandblasting. After the deposition of three layers of Ni/Cu/Ni deposits on PEI substrates, the adhesion of the deposits on PEI substrates was assessed by the cross-cut test with a 3M tape. The Ni/Cu/Ni deposit on the PEI substrate appears silver colored. The surface of the deposit was uniform and dense, with no evidence of defects. The Ni/Cu/Ni multilayer exhibited a poor adhesion with PEI substrates after the cross-cut test (Figure 5a). The cross-scratched deposit was detached from the substrate after peeling the tape. The affected area was in the range of 90~99% of the lattice and the adhesion can be classified as 0B grade. There is no evidence of the Ni/Cu/Ni deposit being removed from the sandblasted PEI substrates after the 3M tape, as shown in Figure 5b. The region between the grids kept intact, exhibiting good adhesion, and were classified as 5B grade. The interfacial adhesion strength between the Ni/Cu/Ni multilayer and PEI composites was remarkably improved by the pretreatment, which was attributed to the mechanical interlocking effect at the interface between the multilayer and the PEI substrates with high surface roughness. However, the Ni/Cu/Ni multilayer on the neat surface of the PEI substrates without the sandblasting process exhibited poor adhesion, which might be attributed to low surface roughness and some contaminations on the surface, such as release agent residues, resulting in poor adhesion between the multilayer and PEI substrates.

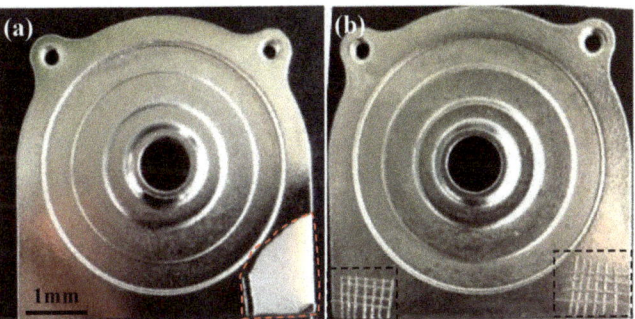

Figure 5. Digital images of Ni/Cu/Ni multilayer deposits on PEI substrates without (**a**) and with (**b**) sandblasting process after cross-cut test by 3M tape.

Figure 6 shows the SEM images and EDS pattern of the surface and cross-section of the initial thin-layer Ni. The surface of the initial Ni layer was rough because of the sandblasted PEI substrate, and some fine microcracks were observed on the surface (Figure 6a). The initial Ni layer consisted of some small aggregates and large particles. The surface of the initial Ni layer on the untreated PEI substrate was smooth; however, a fine microcrack was observed (Figure 6b). The thickness of the initial electroless Ni layer was about 2 μm. The interface between the layer and the substrate was obvious, with no evidence of delamination (Figure 6c). The P content in the initial Ni layer was 5.9 wt%. Generally, electroless Ni–P deposits with less than 5 wt% P have a crystalline structure with an average grain size of approximately 2–15 nm, and the films with more than approximately 10 wt% P show an amorphous phase compared with Ni. Within the intermediate concentration range, nanocrystallite was embedded in an amorphous matrix [18]. Therefore, in this work, this initial Ni layer on the PEI substrate was composed of a mixture of amorphous and nanocrystalline phases.

Figure 7 shows the microstructure and morphology of the surface and fracture of the Ni/Cu/Ni multilayer on the sandblasted PEI substrate. The visual appearance was a continuous and homogeneous layer with a low presence of visible defects (Figure 7a). There is a continuous film. The globular shape of the top of the columnar features can be observed. However, the top-surface of the Ni layer was also composed of some small aggregates with sizes of 2.0~8.0 μm. The surface was relatively rough due to the rough surface of the PEI substrate. The fracture of the multilayer on the PEI composite is displayed in Figure 7b, which was treated by a mechanical cutting machine. The thickness of the top-surface Ni layer was about 6.31 μm. The thickness of the Cu interlayer was about 20 μm. Glass fibers and some holes were observed on the fracture of the PEI substrate. After the sandblasting process, the surface generates a non-uniform surface structure characterized by dimples and furrows. There were no noticeable cracks, and this behavior demonstrated a high-quality composite multilayer. There is no evidence of the delamination between the layers and the deposit and the substrate.

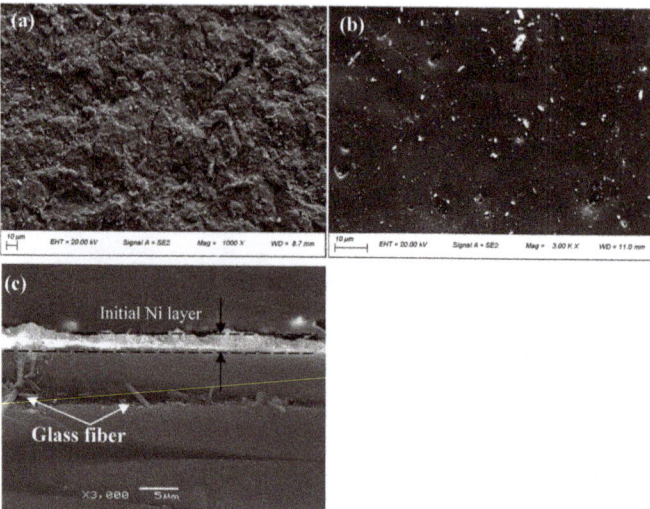

Figure 6. SEM images of surface of alkaline electroless Ni layer on sandblasted PEI substrate (**a**) and on untreated PEI substrate (**b**); SEM image of cross-section of alkaline electroless Ni layer on untreated PEI substrate (**c**).

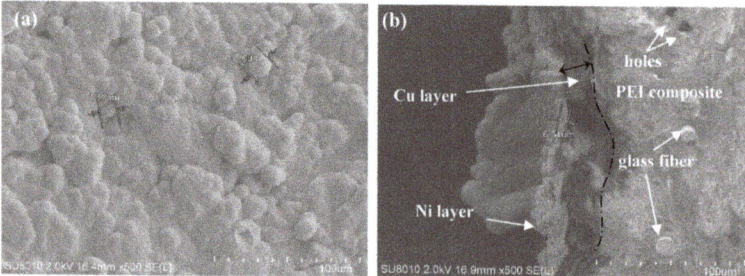

Figure 7. SEM images of top-surface (**a**) and fracture (**b**) of multilayer deposit on PEI substrate after sandblasting.

Generally, colloid palladium was used as a kind of catalyst for the surface metallization of the polymer substrates [39]. The catalytic activity of colloid palladium not only played an important role in the appearance quality of the surface metal layers but also affected the industrialization applications of metallic layers on polymer substrates. The excess $SnCl_2 \cdot 2H_2O$ surrounds palladium to form colloidal palladium. The activation process can be realized by one or two steps. The latter includes the sensitization of a $SnCl_2$ acidic solution and the activation of a $PdCl_2$ acidic solution. Firstly, the substrate was immersed into a solution in order to adsorb Sn cations at the surface of the PEI substrates. Secondly, a batch of samples were immersed in the activation solution containing the catalytic chemical, resulting in the formation of the catalyst species. However, the colloidal palladium solution just needs one step in this work. A commercial solution with a combination of sensitizer and activator contains a palladium complex. The Pd complex activation solution is gray black due to the formation of the $(PdSn_6)Cl_{14}$ complex by the following reaction (6):

$$6SnCl_2 \text{ (excess)} + PdCl_2 = (PdSn_6)Cl_{14} \qquad (6)$$

The colloidal palladium solution will fail. After a period in the beaker, the color of the solution will be changed to yellow. There is no deposit formed in the solution. Figure 8 shows the UV–vis spectra and macrograph of the activation solution in the glass bottles. It can be observed that the colloidal palladium solution shows the gray dark color; however, the color of the colloidal palladium solution became yellow after a period (about 10 days exposed to air). According to the observation of the UV–vis spectra, the chemical state of the ions in the solution was changed after 10 days exposed to air. This is possible due to the oxidization of Sn^{2+} ions, as in the following reaction (7):

$$2SnCl_2 + O_2 + 4HCl = 2SnCl_4 + 2H_2O \quad (7)$$

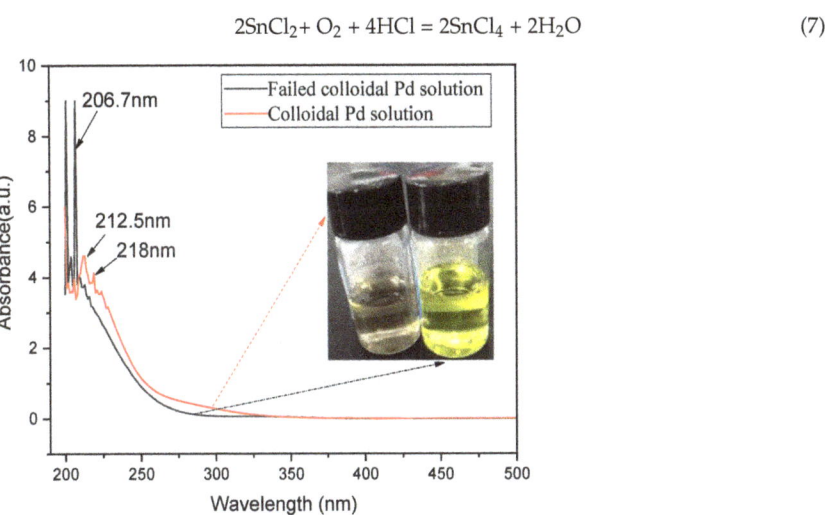

Figure 8. UV–vis spectra and macrograph of activation solution in the glass bottles.

After the PEI substrates were treated and activated by the sensitization/activation process, the substrates were thoroughly rinsed into acceleration solution at the temperature of 70 °C to remove excess hydroxides, oxides, and salts of tin, leaving active catalyst species, such as metallic Pd and $PdSn_x$ nanoparticles [34]. This step is commonly known as acceleration. The accelerator is a strong acid or alkaline solution, which plays an important role in energizing the mixed catalysts. The electroless plating of an initial Ni layer will not occur without the acceleration process. The action and function of the accelerator is to convert hydrolyzed stannous and $PdCl_2$ to active metallic Pd. The surface catalytic activity of the substrate is related to the number of catalytic nuclei of Pd, which depends on the degree of acceleration and the amount of $SnCl_2/PdCl_2$ adsorbed on the surface of PET substrates.

Figure 9 shows the XPS spectra of the top-surface of the sample after the sensitization/activation and acceleration processes. The surface of the PEI substrates treated only with the sensitization/activation solution shows Sn, O, Pd, and C signal peaks. The observed carbon signal was not characteristic of the PEI and is attributed to normal environmental contamination. It can be observed that Sn 4d (26.25 eV), Sn3d (487.76 eV), and Sn 3p (716.4 eV) signals were eliminated and weak after the acceleration process but the Pd and Cl signals were still strong. Therefore, the acceleration process could be helpful to convert hydrolyzed stannous and $PdCl_2$ to active metallic Pd. The Sn^{2+} and Sn^{4+} ions were easily removed from the colloid solution.

Figure 10 shows the high-resolution XPS spectra of the Pd and Sn. The Sn(IV) signal can be observed before acceleration. Before acceleration process, the Pd^0 and a small amount of Pd^{2+} were present, at the same time there are main Sn^{4+} ions at 486.3 eV for several stannous or stannic oxides or hydroxides, and small amounts of Sn^0 at 485.5 eV in the solution. After the acceleration process, both Sn and Sn(IV) species were present, and colloidal Pd was mainly present in Pd^0 metallic state. It can be indicated that the primary Pd species on the surface of the PEI substrate are in the metallic state before and after acceleration, although a fraction of the Pd is oxidized during the sensitization/activation and acceleration processes, which could be attributed to the sensitization/activation solution being used for a period. After the acceleration process, the intensity of the Sn 3d5/2 signal peak decreases dramatically, and two nearly equal reflections were observed with binding energies of 485.2 eV and 486.8 eV (Figure 10d). The reflection at a relatively low binding energy is attributed to metallic Sn species, while the reflection at a high binding energy is due to the remaining stannic hydroxides. Thus, it is assumed that only the metallic Sn is associated with the active catalytic center. After the acceleration, lots of Sn species have been removed. The highly active catalyst of Pd^0 induces Ni deposition instantaneously by immersion in an electroless plating Ni bath.

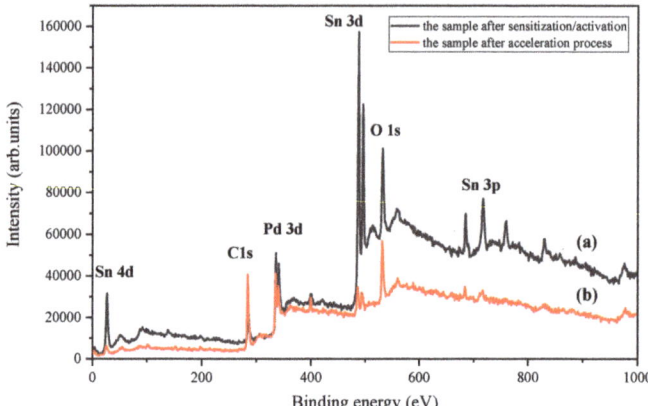

Figure 9. XPS spectra of top-surface of the sample after sensitization/activation process (**a**) and after acceleration process (**b**).

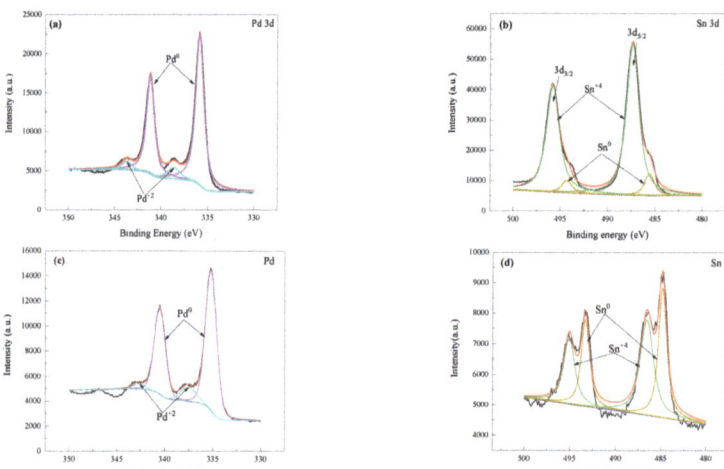

Figure 10. High-resolution XPS spectra of Pd 3d and Sn 3d before and after acceleration process. XPS of Pd 3d (**a**) and Sn 3d (**b**) before acceleration process, and XPS of Pd 3d (**c**) and Sn 3d (**d**) after acceleration process.

The initial nickel solution is an alkaline solution produced from electroless plating. This solution is strongly alkaline (pH > 12). This bath produces a binary nickel–phosphorous (Ni-P) alloy deposit. In an alkaline bath based on the dehydrogenation of the reducing agent, the deposition mechanism of the electroless Ni-P coating involved the following chemical reactions [16,18]:

$$H_2PO_2^- + 2OH^- \rightarrow H^- + H_2PO_3^- + H_2O \tag{8}$$

$$Ni^{2+} + 2H^- \rightarrow Ni^0 + H_2\uparrow \qquad (9)$$

$$H_2PO_2^- + H^- \rightarrow 2OH^- + P + 1/2H_2\uparrow \qquad (10)$$

$$H_2O + H^- \rightarrow OH^- + H_2\uparrow \qquad (11)$$

According to Equations (9)–(11), the formation of hydrogen bubbles in the solution was from the hydride ions and water. After a standard acceleration time, the masking salt and the reactive Pd or metallic PdSn$_x$ were exposed, which then initiates the deposition of Ni, forming a conductive Ni layer for the next step of electroplating. The complete Ni layer was deposited on the PEI substrate, as displayed in Figure 11a. However, the acceleration time will affect the deposition of Ni for electroless plating (see Figure 11b,c). In the case of insufficient acceleration time, the colloid reacts with the strong acid insufficiently, resulting in incomplete exposure of the active Pd. Therefore, only exposed active Pd^0 atoms or metallic PdSn$_x$ alloys can induce the reduction of Ni^{2+} ions and the formation of the initial Ni layer. However, under excessively accelerated conditions, prolonged exposure to acidic media may cause the entire colloidal shell (including the portion locked to the substrate surface) to react, followed by the leaching of the Pd core, resulting in a loss of catalytic activity; furthermore, excessively prolonged acceleration treatment reduces catalytic activity. During the long acceleration step, a part of Pd may be oxidized [36]. At the same time, stirring the solution and air may dislodge the Pd nuclei from the substrate surface. Therefore, the appearance quality of the Ni deposit on the PEI substrate was poor, showing a skip plating defect.

Figure 11. Digital macrographs of PEI samples: (**a**) standard condition (4 min), (**b**) short acceleration time (2 min), (**c**) excessive acceleration time (10 min).

As displayed in Figure 12, colloid Pd mainly comprised Pd as the activation core and Sn and Cl as a protective shell core [38]. Firstly, the colloidal nucleus selectively adsorbed the Sn^{2+} and Sn^{4+} ions in the solution to form a dense layer. The colloidal Pd-Sn nanoparticles were formed with a Pd nucleus as the core, and Sn^{n+} as the shell core. Then, the colloidal Pd-Sn nanoparticles with a positive charge would absorb the Cl$^-$ ions with a negative charge for neutralization. A ring of encirclement was generated around the colloidal Pd-Sn nanoparticles in order to protect the Pd core. Therefore, the electrically neutral colloid Pd was obtained in the form of Pd$_x$Sn$_y$Cl$_z$ or {Pd$_m$·2Sn^{n+}·2(n − x)Cl$^-$}·2xCl$^-$ complexes [40]. After acceleration, the chemical nature of the adsorbed species was also addressed in previous research [29,31]. Cohen and Meek [41] discovered that Sn^{4+} in the form of stannic hydroxide was found before the acceleration process, and that both Sn(OH)$_4$ and Pd-Sn alloys were present after acceleration. Osaka et al. [42] examined catalyzed surfaces and reported that a Pd–Sn alloy was formed with the removal of Sn^{4+} ions during acceleration. Burrell et al. [36] reported that only Sn^{4+} was found before acceleration, and both metallic Sn and Sn^{4+} were present after acceleration; furthermore, he found that metallic Pd^0 was primarily present before and after the acceleration process. Pd^{2+} species from a PdCl$_2$ solution were attached on the PEI surface and, subsequently, Pd^{2+} species were reduced into Pd^0. The Pd^0 catalyst allowed an instantaneous initiation of the Ni deposit by immersion in a plating bath. After the standard acceleration time, the masking salts are leached away and exposed the active Pd or the metallic PdSn$_x$, which is then able to initiate the deposition of the Ni layer, forming a conductive layer. However, if the acceleration time after the sensitization/activation process is too long or short, it could influence the deposition of the electroless Ni layer. In insufficient acceleration conditions, the colloid does not fully react with the strong acid and results in an incomplete exposure of the active Pd nanoparticles. Therefore, only the exposed active Pd atom or the metallic PdSn$_x$ alloy nanoparticles can induce the reduction of Ni^{2+} ions and form the initial Ni layer. However, in the excessive acceleration condition, prolonged exposure to the acid media may lead to the colloid shells reaction, which then leach out of the Pd nuclei, resulting in some deterioration of catalytic activity through prolonged treatment [43]. During acceleration, a fraction of the Pd nanoparticles might be oxidized after a period of time [36]. At the same time, the Pd^0 nuclei might be detached from the substrate surface by stirring the solution and air-blowing in the electrolyte. The excessive acceleration condition could result in the non-uniform and discontinuous layer deposited on the surface of the PEI substrates.

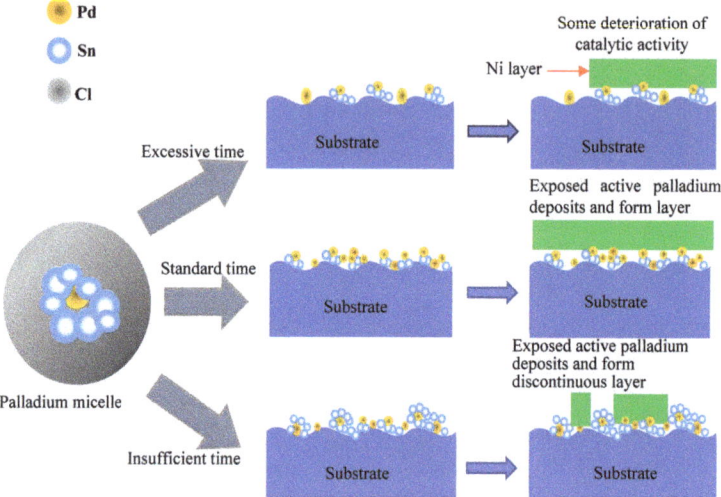

Figure 12. Illustration of acceleration of Pd micelle and formation of initial Ni layer.

4. Conclusions

Ni/Cu/Ni multilayer metallization on PEI resin reinforced with glass fibers was successfully made by a two-step metallization process, including sandblasting and activation/acceleration. However, the pretreatment processes have an influence on the appearance quality and adhesion of the Ni/Cu/Ni multilayer deposits on the PEI substrates. The surface roughening process, namely sandblasting, influenced the adhesion and appearance quality of the deposits on the PEI substrates. The average surface roughness of the substrates with and without the sandblasting process were 117 nm and 2125 nm, respectively. The surface roughness of the PEI substrate significantly increased after sandblasting, which could improve the adhesion of the metallic multilayer. The pretreatment processes of sensitization/activation and acceleration affected the appearance quality of the deposit on the PEI substrate. After the sensitization/activation process, Pd^0 and a small amount of Pd^{2+} were present, and there were lots of Sn^{4+} ions for a few stannous or stannic oxides or hydroxides. However, the metallic state was the primary Pd species on the substrate after acceleration. The acceleration had an effect on the deposition of Ni for electroless plating. After the standard acceleration time, the initial electroless continuous Ni layer was formed on the surface of the PEI substrates as a conductive layer for the next step of the electroplating process.

Author Contributions: W.W. planned, wrote, and supervised this work; X.X. and D.X. prepared the multilayer films and co-wrote and edited this paper; W.W. supervised Masters student D.X.; X.X., D.X., J.H. and G.H. performed some experiments and tests; Y.Z., W.W. and P.J. performed and analyzed characterization of materials and samples; K.L. and L.T. conceived the rest of the experiments and provided experimental platform and three months of internship experience for Masters students. All authors have read and agreed to the published version of the manuscript.

Funding: This research was funded by Jiaqi Huang acknowledges funding from the Postgraduate Research & Practice Innovation Program of Jiangsu Province (Grant no. SJCX22_1429).

Institutional Review Board Statement: Not applicable.

Informed Consent Statement: Not applicable.

Data Availability Statement: Not applicable.

Acknowledgments: The authors thank Ying Wang and Jun Jiang from the Analysis and Testing Center of Changzhou University for discussion and helping in the surface roughness and XPS measurement, respectively.

Conflicts of Interest: The authors declare no conflict of interest.

References

1. Viscusi, A.; Durante, M.; Astarita, A.; Boccarusso, L.; Carrino, L.; Perna, A.S. Experimental Evaluation of Metallic Coating on Polymer by Cold Spray. *Procedia Manuf.* **2020**, *47*, 761–765. [CrossRef]
2. Esfahani, R.N.; Shuttleworth, M.P.; Doychinov, V.; Wilkinson, N.J.; Hinton, J.; Jones, T.D.A.; Ryspayeva, A.; Robertson, I.D.; Marques-Hueso, J.; Desmulliez, M.P.Y.; et al. Light based synthesis of metallic nanoparticles on surface-modified 3D printed substrates for high performance electronic systems. *Addit. Manufact.* **2020**, *34*, 101367. [CrossRef]
3. Ding, S.; Zou, B.; Wang, P.; Ding, H. Effects of nozzle temperature and building orientation on mechanical properties and microstructure of PEEK and PEI printed by 3D-FDM. *Polym. Test.* **2019**, *78*, 105948. [CrossRef]
4. Polyakov, I.V.; Vaganov, G.V.; Yudin, V.E.; Smirnova, N.V.; Ivan'kova, E.M.; Popova, E.N. Study of polyetherimide and its nanocomposite 3D printed samples for biomedical application. *Polym. Sci. Ser. A* **2020**, *62*, 337–342. [CrossRef]
5. Melentiev, R.; Yu, N.; Lubineau, G. Polymer metallization via cold spray additive manufacturing: A review of process control, coating qualities, and prospective applications. *Addit. Manuf.* **2021**, *48*, 102459. [CrossRef]
6. Kouicem, M.M.; Tomasella, E.; Bousquet, A.; Batisse, N.; Monier, G.; Robert-Goumet, C.; Dubost, L. An investigation of adhesion mechanisms between plasma-treated PMMA support and aluminum thin films deposited by PVD. *Appl. Surf. Sci* **2021**, *564*, 150322. [CrossRef]
7. Ferreira, A.A.; Silva, F.J.G.; Pinto, A.G.; Sousa, V.F.C. Characterization of Thin Chromium Coatings Produced by PVD Sputtering for Optical Applications. *Coatings* **2021**, *11*, 215. [CrossRef]
8. Duguet, T.; Senocq, F.; Laffont, L.; Vahlas, C. Metallization of polymer composites by metalorganic chemical vapor deposition of Cu: Surface functionalization driven films characteristics. *Surf. Coat. Technol.* **2013**, *230*, 254–259. [CrossRef]
9. Addou, F.; Duguet, T.; Ledru, Y.; Mesnier, D.; Vahlas, C. Engineering Copper Adhesion on Poly-Epoxy Surfaces Allows One-Pot Metallization of Polymer Composite Telecommunication Waveguides. *Coatings* **2021**, *11*, 50. [CrossRef]
10. Gonzalez, R.; Ashrafizadeh, H.; Lopera, A.; Mertiny, P.; McDonald, A. A Review of Thermal Spray Metallization of Polymer-Based Structures. *J. Therm. Spray Technol.* **2016**, *25*, 897–919. [CrossRef]
11. Parmar, H.; Tucci, F.; Carlone, P.; Sudarshan, T.S. Metallisation of polymers and polymer matrix composites by cold spray: State of the art and research perspectives. *Int. Mater. Rev.* **2021**, *67*, 385–409. [CrossRef]
12. Shacham-Diamand, Y.; Osaka, T.; Okinaka, Y.; Sugiyama, A.; Dubin, V. 30 years of electroless plating for semiconductor and polymer micro-systems. *Microelectron. Eng.* **2015**, *132*, 35–45. [CrossRef]
13. Ghosh, S. Electroless copper deposition: A critical review. *Thin Solid Films* **2019**, *669*, 641–658. [CrossRef]
14. Huang, J.; Gui, C.; Ma, H.; Li, P.; Wu, W.; Chen, Z. Surface metallization of PET sheet: Fabrication of Pd nanoparticle/polymer brush to catalyze electroless nickel plating. *Compos. Sci. Technol.* **2021**, *202*, 108547. [CrossRef]
15. Dupenne, D.; Lonjon, A.; Dantras, E.; Pierré, T.; Lubineau, M.; Lacabanne, C. Carbon fiber reinforced polymer metallization via a conductive silver nanowires polyurethane coating for electromagnetic shielding. *J. Appl. Polym. Sci.* **2020**, *138*, 50146. [CrossRef]
16. Wu, W.P.; Liu, J.W.; Miao, N.M.; Jiang, J.J.; Zhang, Y.; Zhang, L.; Yuan, N.Y.; Wang, Q.Q.; Tang, L.X. Influence of thiourea on electroless Ni-P films on silicon substrates. *J. Mater. Sci. Mater. Electron.* **2019**, *30*, 7717–7724. [CrossRef]
17. Wu, W.; Wang, X.; Xie, D.; Zhang, Y.; Liu, J. Corrosion failure analysis of Ni-P film of aircraft fire detector components. *Eng. Fail. Anal.* **2020**, *111*, 104497. [CrossRef]
18. Wu, W.-P.; Jiang, J.-J. Effect of plating temperature on electroless amorphous Ni–P film on Si wafers in an alkaline bath solution. *Appl. Nanosci.* **2017**, *7*, 325–333. [CrossRef]
19. Dixit, N.K.; Srivastava, R.; Narain, R. Improving surface roughness of the 3D printed part using electroless plating. *Proc. Inst. Mech. Eng. Part L J. Mater. Des. Appl.* **2019**, *233*, 942–944. [CrossRef]

20. Huang, J.; Wu, W.; Zhang, R.; Lu, G.; Chen, B.; Chen, Z.; Gui, C. Novel electrode material using electroless nickel plating for triboelectric nanogenerator: Study of the relationship between electrostatic-charge density and strain in dielectric material. *Nano Energy* **2022**, *92*, 106734. [CrossRef]
21. Marques-Hueso, J.; Jones, T.D.A.; Watson, D.E.; Ryspayeva, A.; Esfahani, M.N.; Shuttleworth, M.P.; Harris, R.A.; Kay, R.W.; Desmulliez, M.P.Y. A rapid photopatterning method for selective plating of 2D and 3D microcircuitry on polyetherimide. *Adv. Funct. Mater.* **2018**, *28*, 1704451–1704459. [CrossRef]
22. Marline, C.; Maurice, R.; Yves, G. Direct Ni electroless metallization of poly(etherimide) without using palladium as a catalyst. *Trans. Mater. Heat Treat. Proc. 14th IFHTSE Congr.* **2004**, *25*, 1106–1111.
23. Jones, T.D.A.; Ryspayeva, A.; Esfahani, M.N.; Shuttleworth, M.P.; Harris, R.A.; Kay, R.W.; Desmulliez, M.P.Y.; Marques-Hueso, J. Direct metallisation of polyetherimide substrates by activation with different metals. *Surf. Coat. Technol.* **2019**, *360*, 285–286. [CrossRef]
24. Alodan, M.A. Metallizing Polyetherimide Resin Reinforced with Glass Fibers. *J. King Saud Univ. Eng. Sci.* **2005**, *17*, 251–259. [CrossRef]
25. Zhang, F.-T.; Xu, L.; Chen, J.-H.; Zhao, B.; Fu, X.-Z.; Sun, R.; Chen, Q.; Wong, C.-P. Electroless Deposition Metals on Poly(dimethylsiloxane) with Strong Adhesion as Flexible and Stretchable Conductive Materials. *ACS Appl. Mater. Interfaces* **2018**, *10*, 2075–2082. [CrossRef] [PubMed]
26. Wu, W.; Xie, D.; Huang, J.; Wang, Q.; Chen, Q.; Huang, J. Adhesion enhancement for nickel layer deposited on carbon fiber reinforced polymer (CFRP) composites by pretreatment processes for lightning strike. *J. Adhes.* **2022**, 1–24. [CrossRef]
27. Fischer, A.J.; Meister, S.; Drummer, D. Effect of fillers on the metallization of laser-structured polymer parts. *J. Polym. Eng.* **2017**, *37*, 151–161. [CrossRef]
28. Gebauer, J.; Burkhardt, M.; Franke, V.; Lasagni, A.F. On the Ablation Behavior of Carbon Fiber-Reinforced Plastics during Laser Surface Treatment Using Pulsed Lasers. *Materials* **2020**, *13*, 5682. [CrossRef]
29. Li, J.F.; Lin, P.T.; Sun, H.R.; Li, X.X.; Sang, J.; Wang, X.G.; Li, Q.; Jin, Y.; Zhang, L.G. Selective metallization on CFRP composites by laser radiation and electroless plating. *J. Phys. Conf. Ser.* **2021**, *2101*, 012048. [CrossRef]
30. Puliyalil, H.; Filipič, G.; Cvelbar, U. Chapter 9-Selective plasma etching of polymers and polymer matrix composites. In *Non-Thermal Plasma Technology for Polymeric Materials-Applications in Composites, Nanostructured Materials and Biomedical Fields*; Elsevier: Amsterdam, The Netherlands, 2019; pp. 241–259.
31. Rao, C.H.; Kothuru, A.; Singh, A.P.; Varaprasad, B.K.S.V.L.; Goel, S. Plasma Treatment and Copper Metallization for Reliable Plated-Through-Holes in Microwave PCBs for Space Electronic Packaging. *IEEE Trans. Compon. Packag. Manuf. Technol.* **2020**, *10*, 1921–1928. [CrossRef]
32. Xie, D.; Wu, W.; Huang, J.; Wang, X.; Zhang, Y.; Wang, Z.; Jiang, P.; Tang, L. Effect of electrodeposited Cu interlayer thickness on characterizations and adhesion force of Ni/Cu/Ni coatings on polyetherimide composite substrates. *Int. J. Adhes. Adhes.* **2021**. Available online: https://www.researchgate.net/publication/355181252 (accessed on 8 July 2022).
33. Ramaswamy, K.; O'Higgins, R.M.; Kadiyala, A.K.; McCarthy, M.A.; McCarthy, C.T. Evaluation of grit-blasting as a pre-treatment for carbon-fibre thermoplastic composite to aluminium bonded joints tested at static and dynamic loading rates. *Compos. Part B Eng.* **2020**, *185*, 107765. [CrossRef]
34. Radoeva, M.; Monev, M.; Ivanov, I.; Georgiev, G.; Radoev, B. Adhesion improvement of electroless copper coatings by polymer additives. *Colloids Surf. A Physicochem. Eng. Asp.* **2014**, *460*, 441–447. [CrossRef]
35. Guo, R.; Yin, G.; Sha, X.; Wei, L.; Zhao, Q. Effect of surface modification on the adhesion enhancement of electrolessly deposited Ag-PTFE antibacterial composite coatings to polymer substrates. *Mater. Lett.* **2015**, *143*, 256–260. [CrossRef]
36. Burrell, M.C.; Smith, G.A.; Chera, J.J. Characterization of $PdCl_2/SnCl_2$ metallization catalysts on a polyetherimide surface by XPS and RBS. *Surf. Interf. Analy* **1988**, *11*, 160–164. [CrossRef]
37. Touyeras, F.; Hihn, J.; Bourgoin, X.; Jacques, B.; Hallez, L.; Branger, V. Effects of ultrasonic irradiation on the properties of coatings obtained by electroless plating and electro plating. *Ultrason. Sonochem.* **2005**, *12*, 13–19. [CrossRef]
38. Nicolas-Debarnot, D.; Pascu, M.; Vasile, C.; Poncin-Epaillard, F. Influence of the polymer pre-treatment before its electroless metallization. *Surf. Coat. Technol.* **2006**, *200*, 4257–4265. [CrossRef]
39. Zhao, Y.; Zhan, L.; Tian, J.; Nie, S.; Ning, Z. Enhanced electrocatalytic oxidation of methanol on Pd/polypyrrole–graphene in alkaline medium. *Electrochim. Acta* **2011**, *56*, 1967–1972. [CrossRef]
40. Shuai, H.; Jian, W.; Yu, H.L.; Tang, J.B.; Zhang, X. Microstructure characterization and formation mechanism of colloid palladium for activation treatment on the surface of PPTA fibers. *Appl. Surf. Sci.* **2020**, *516*, 146134. [CrossRef]
41. Cohen, R.; Meek, R. The chemistry of palladium—tin colloid sensitizing processes. *J. Colloid Interface Sci.* **1976**, *55*, 156–162. [CrossRef]
42. Osaka, T. A Study on activation and acceleration by mixed $PdCl_2/SnCl_2$ catalysts for electroless metal deposition. *J. Electrochem. Soc.* **1980**, *127*, 390–394. [CrossRef]
43. Rantell, A.; Holtzman, A. The Role of Accelerators Prior to Electroless Plating of ABS Plastic. *Trans. IMF* **1974**, *52*, 31–38. [CrossRef]

Article

Cause Analysis and Solution of Premature Fracture of Suspension Rod in Metro Gear Box

Wenming Liu [1,2], Zhiqiang Xu [1], Hongmei Liu [1,2,3] and Xuedong Liu [1,2,*]

1 School of Mechanical Engineering and Rail Transit, Changzhou University, Changzhou 213164, China
2 Jiangsu Key Laboratory of Green Process Equipment, Changzhou 213164, China
3 Jiangsu Meilan Chemical Co., Ltd., Taizhou 225300, China
* Correspondence: xdliu_65@126.com

Abstract: Through the appearance observation of suspension rod in the metro gearbox, macro and micro observation of the fracture and quantitative analysis of the fracture, combined with the metallographic and hardness examination results of the boom, the finite element model was established and the force analysis of suspension rod was carried out to explore the causes of the fracture of the gearbox boom. The results show that the nature of suspension rod fracture is fatigue. The cause of its fatigue fracture is related to the low fatigue tolerance for booms in metro operation, and the surface shallow decarburization plays a role in promoting the fatigue fracture of suspension rod. The life of fatigue crack growth in the boom is 819 stations (or 1210 km), and the fatigue initiation life is 522,452 km.

Keywords: boom; fracture; finite element analysis

1. Introduction

The gearbox is a key component in metro vehicles and its main function is to transmit the power output from the traction motor to the wheel pairs to drive the vehicle [1]. The gearbox is connected to the lifting base on the bogie frame using a spreader bar. The spreader bar device carries the loads that occur during the operation of the gearbox, including those caused by traction and braking, vibration shocks and loads caused by short circuits in the traction motor [2,3].

The gearbox boom is a traction device installed on the bogie of a rail vehicle to improve the efficiency of traction and braking force transfer between the locomotive and the bogie [2]. The installation of the boom enables the gearbox to withstand the vibration impact from the wheelset during vehicle operation. If there is a relative movement between the wheelset and the bogie frame during operation, the rubber nodes on the boom device will enable the gearbox mounted on the wheelset to move consistent with the displacement of the wheelset so as not to affect other components [4–6]. The boom plays a key role in the safe operation of the gearbox and the vehicle.

F. CURÀ et al. [7] carried out numerical simulations using the three-dimensional extended finite element method to investigate the relationship between the rim crack expansion path and the thickness of the web during bending fatigue failure of thin-sided gears, showing that the thickness of the web affects the rim crack expansion state and the form of failure. Giovanni Meneghetti [8] and others carried out single tooth fatigue tests on gears with hardened tooth surfaces, predicted their fatigue life based on the test results and compared it with that of unhardened gears and found that the sample-based method had the same accuracy as the baseline gear-based method, provided that the material notch sensitivity factor was properly calibrated. Edoardo Conrado et al. [9] compared the flexural fatigue strength of carburized gears with that of nitrided gears. Single tooth flexural fatigue (STBF) tests were carried out on gears of the two different processes, and

S-N curves were obtained for both processes to estimate their fatigue limits. The S-N curves for both processes were obtained and the fatigue limits were estimated.

In this paper, the appearance of the fractured boom was inspected, and the fracture was analyzed by macroscopic and microscopic observation, chemical composition check and mechanical properties evaluation, etc. The fracture mechanism of the boom was determined, and the causes of its production were analyzed qualitatively and quantitatively, to provide analytical methods and references for avoiding the recurrence of such incidents.

2. Materials and Processing Methods

A line metro was inspected and three pieces of gearbox boom were found to be fractured in different units of the vehicle, as shown in Figure 1. The boom material is C45E4 steel, processed in the following steps: raw materials-forging-heat treatment-shot blasting (Φ2 mm steel balls, time about 25–30 min)-magnetic particle testing-primer coating-Machining-Galvanizing (threaded rod section)-Inventory inspection-Finish coating (paint thickness required $\geq$120 μm). The heat treatment system is as follows: normalizing temperature 870–890 °C, air cooling, quenching temperature 840–860 °C, water cooling, tempering temperature 530–600 °C, air cooling. The technical requirements for the mechanical properties of the boom material are shown in Table 1.

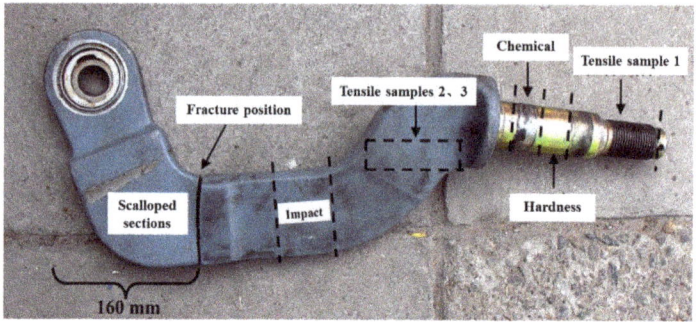

(a)

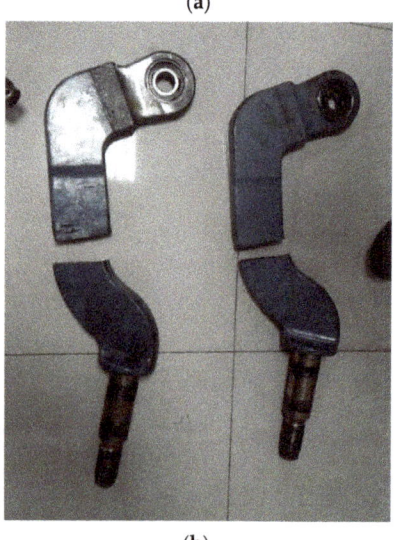

(b)

Figure 1. Appearance of booms. (**a**) Sampling areas; (**b**) Appearance of fractured booms.

Table 1. Technical requirements for the mechanical properties of the boom.

Yield Strength/MPa	Tensile Strength/MPa	Elongation/%	Area Reduction in Tensile Test/%	Impact Value/J/cm^2	Hardness/HB
≥490	≥690	≥17	≥45	≥78	201~269

3. Result Analysis

3.1. Macroscopic Observation

The fracture occurred at the bend transition of the boom structure [10]. Typical fatigue beach marks and extended radial marks are visible in the fracture, and the origin of the fracture can be judged from the direction of convergence of the prisms to be located on the lateral side of the boom width, and the crack is extended along the boom width. From the origin of the fracture to about 24 mm from the origin of the fracture, the fracture surface in the early-term propagation area is flat and the fracture fatigue beach mark is not obvious. From 24 mm to 47 mm from the origin of the fracture, the obvious fatigue beach mark feature is visible in the section. From 47 mm to 82 mm from the origin of the fracture, the fatigue feature is visible in the width of about 4 mm on both sides, and the section in the middle of the width direction is a rough tearing area. From 82 mm to 90 mm from the origin of the fracture, the obvious fatigue beach mark feature is visible in the section. The obvious fatigue beach mark can be seen from 82 mm to 90 mm from the origin of the fracture. Finally, after 90 mm from the origin of the fracture, the section is rough and is a transient fracture zone; see Figures 2 and 3.

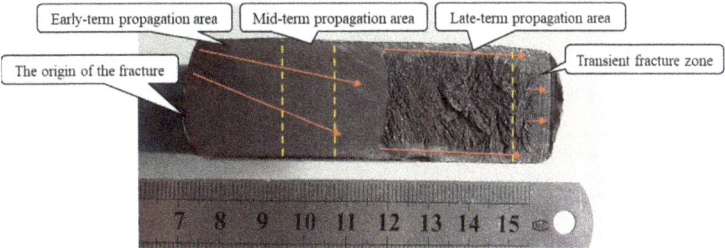

Figure 2. Macroscopic view of boom fracture (Unit: mm).

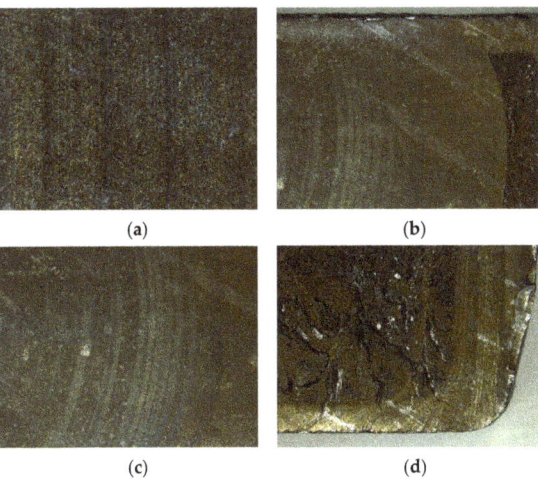

Figure 3. Boom fracture body view. (a) Pre-extension region; (b) Mid-term propagation area; (c) Late-term propagation area; (d) Transient fracture zone.

3.2. SEM Images of Fracture Surfaces

The boom fracture was ultrasonically cleaned with acetone and then analyzed in a scanning electron microscope for observation [11–13]. The origin of the boom fracture is located on the surface, no metallurgical defects are seen, fatigue beach marks and a large number of fine fatigue striations are visible during the cracking process, and the width of the fatigue beach marks from 24 mm to the origin of the fracture to about 90 mm from the origin of the fracture is about 0.1~0.14 mm, and the transient fracture area is characterized by dimples, see Figure 4.

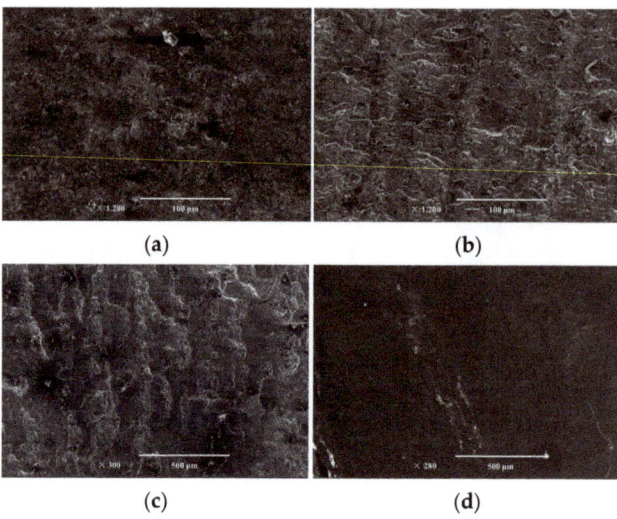

Figure 4. Boom fracture microscopy. (**a**) Early-term propagation area; (**b**) Mid-term propagation area fatigue beach marks and fatigue striations; (**c**) Late-term propagation area fatigue beach marks and fatigue striations; (**d**) Characteristics of the transient fracture zone of the dimple.

3.3. Metallography Analysis

A longitudinal metallographic specimen was taken from the origin of the boom fracture and sent for inspection [14,15], with the metallographic grinding surface perpendicular to the fracture surface, and the macroscopic morphology after etching with 4% nitric acid in alcohol is shown in Figure 5.

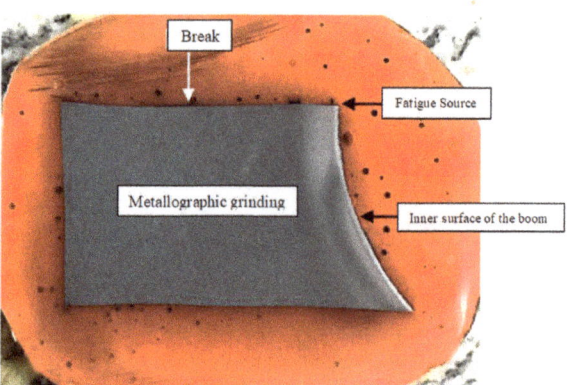

Figure 5. The macroscopic metallographic surface of the origin of the boom fracture sent for inspection.

It can be observed from Figure 6 that in the unetched high magnification morphology of the origin of the boom fracture sent for inspection, the area can be seen containing fine cracks and small surface pits, and no obvious coarse inclusions and other defects exist. The microstructure of the origin of the fracture after etching is fully decarburized, with a thickness of approximately 0.25 mm, and a small amount of plastic deformation in the local area below the origin of the fracture near the inner surface of the boom.

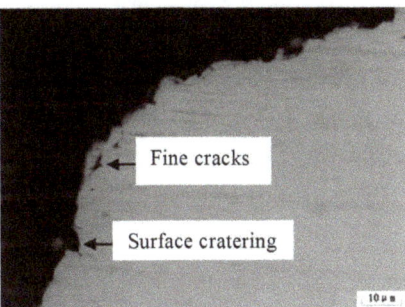

Figure 6. Unetched high magnification morphology of the origin of the boom fracture sent for inspection.

When the inner surface of the boom on the fatigue source side of the fracture was observed, cracks (Crack 1, Crack 2 and Crack 3) were found extending from the inner surface of the boom in a vertical direction to the interior, with crack lengths of approximately 0.13 mm, 1.95 mm and 1 mm respectively, and distances to the fracture of approximately 0.5 mm, 5 mm and 10.5 mm respectively, as shown in Figure 7. It is also seen that the thickness of the fully decarburized layer near the inner surface of the boom has an overall decreasing trend from the origin of the fracture downwards (0.25–0.05 mm), but local areas can be thick or narrow.

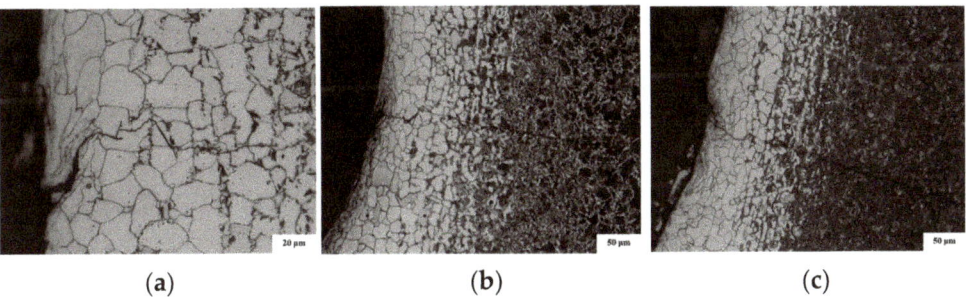

Figure 7. Crack characteristics of the sending boom near the inner surface (below the origin of the fracture). (**a**) Crack 1; (**b**) Crack 2; (**c**) Crack 3.

Near the origin of the fracture, five different microstructure areas (I, II, III, IV and V) can be seen from the inner surface of the boom to the core of the boom, as shown in Figure 8, with micro-Vickers hardness values of approximately 137.0 HV0.2 (Microstructure I), 246.0 HV0.2 (Microstructure II), 265.0 HV0.2 (Microstructure III), 250.0 HV0.2 (Microstructure IV), 252.0 HV0.2 (Microstructure V), 250.0 HV0.2 (Microstructure IV) and 252.0 HV0.2 (Microstructure V). The microstructure of the inner surface of the spreader bar to the core gradually transitions from fully decarburized (microstructure I) to tempered sorbite and white ferrite with a reticulated distribution, with some of the ferrite developing needle-like into the grain (microstructure II), and then to tempered sorbite that maintains the martensitic phase (microstructure III). The white ferrite in microstructure IV is slightly more severe than in microstructure V. Microstructures I and II are known as the transition zone. Below the origin of the fracture, the transition zone from the inner surface of the boom to the core of the boom becomes progressively narrower. The microstructure of the fractured boom is finer-grained, with microstructure I (fully decarburized) having a grain size of 8.5 and microstructures II to V having a grain size of 9.

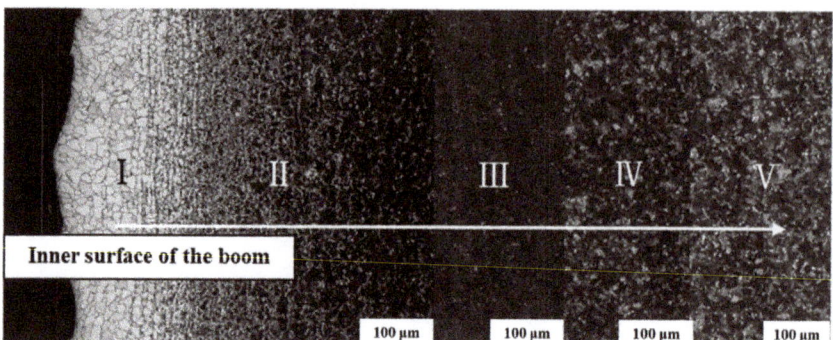

Figure 8. Microstructure of the boom near the fatigue source from the inner surface of the boom to the core.

The non-metallic inclusions in the gearbox boom were assessed in accordance with method A of GB/T 10561-2005 "Method for the determination of non-metallic inclusions in steel" and the non-metallic inclusions in type A (sulfides), B (alumina), C (silicates) and D (spherical oxides) of the fracture boom were recognized as coarse inclusions, the level is less than 0.5 [16].

3.4. Chemical Composition Test Results

The chemical composition of the fracture boom was tested according to the "Gearbox Boom Procurement Standard Book" provided by the client, and the test method was GB/T 4336-2002 "Carbon steel and low and medium alloy spark source yard emission spectroscopy method (conventional method)". The results are shown in Table 2. The chemical composition of the fracture boom sent for testing met the technical requirements provided by the commissioner.

Table 2. Test results for the chemical composition of gearbox boom sent for inspection (mass, %).

Ingredient	C	Si	Mn	P	S
Composition	0.462	0.230	0.667	0.006	0.001

3.5. Mechanical Performance Testing

The tensile test, hardness test and impact test were carried out on the boom test according to the relevant standards respectively [17–19]. Hardness tests were carried out on the surface and core hardness of the boom. Specimens were taken in the positions shown in Figure 1 and subjected to tensile and impact tests, and the results are shown in Table 3. It can be seen that the yield strength of the fracture boom sent for inspection is lower than the technical requirements provided by the commissioner, while the tensile strength, axial elongation and area reduction in the tensile test meet the relevant technical requirements. The average hardness of the surface decarburization layer is 167 HB, which is much lower than the technical requirements (201~269 HB). The average hardness of the core is 230 HB, which meets the technical requirements (201~269 HB). The impact performance of the boom meets the technical requirements provided by the commissioning party.

Table 3. Gearbox boom mechanical properties test results.

Specimen	Tensile Test				Brinell Hardness		Impact Value (J)
	Yield Strength (MPa)	Tensile Strength (MPa)	Axial Elongation (%)	Area Reduction in Tensile Test (%)	Surface Decarburization	Core Hardness	
AVG	476	753.3	22.8	67.3	167	230	83.3
STD	7.21	6.66	1.44	0.58	7.94	4.00	4.62

4. Finite Element Analysis

In the gearbox transmission system, the gearbox is suspended from the bogie frame by means of a boom connection. The boom is not only subjected to tensile and compressive loads during the use of the gearbox but also to various impact loads during operation [20]. In order to analyze whether the strength of the boom meets the design requirements and analyze the stress distribution of the boom and make suggestions and recommendations for design improvements, a calculation of the static strength of the boom is required.

4.1. Model Simplification

In this paper, the finite element calculations are carried out using Pro/Engineer software for 3D modeling, and the model is imported into the finite element software for finite element analysis calculations.

The boom is first analyzed under different operating conditions to derive its maximum force load and thus its stress distribution under this load. The boom is subjected to a force of F_r, the pinion end is driven by the motor and the torque applied is Md. The torque at the large gear end is $Md \times i$, as can be seen from the transmission relationship between the large and small gears.

The input parameters of the motor, without regard to vibration, are as follows:

- The rated torque of the motor is: 955 Nm;
- The maximum traction (braking) torque of the motor is: 1361 Nm;
- Short-circuit torque of the motor is: 8000 Nm;
- Transmission ratio i: 7.69;
- L is the distance from the boom centerline to the axle centerline, 421.68 mm;
- Then the torque balance gives $F_r = Md \times (1+i)/L$;
- At rated operation: $F_{r1} = 955 \times (7.69 + 1)/421.68 = 19.68$ kN;
- At start-up: $F_{r2} = 1361 \times (7.69 + 1)/421.68 = 28.05$ kN.

The input parameters for the case where vibration (vertical) is considered are as follows.

- W is the mass of the case, 131 kg;
- $Wp + Wc/2$ is the pinion weight and half coupling mass, 20.6 kg;
- W_r is the boom mass, 15.7 kg;
- Maximum vertical vibration acceleration at the boom, ±15 g;
- Vibration force: $Fa = (W/3 + Wp + Wc/2 + Wr) \times 15$ g = 11.77 kN.

Boom force load:

- Rated working condition: $F1 = Fr1 + Fa = 31.45$ kN;
- Start-up condition: $F2 = Fr2 + Fa = 39.82$ kN.

The boom finite element model is shown in Figure 9.

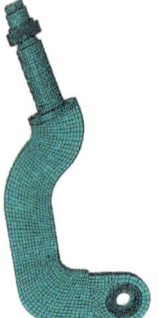

Figure 9. Boom finite element model.

4.2. Calculation Results

The stress distribution of the boom under the two working conditions is obtained by calculation, respectively, see Figure 10 (stresses in MPa).

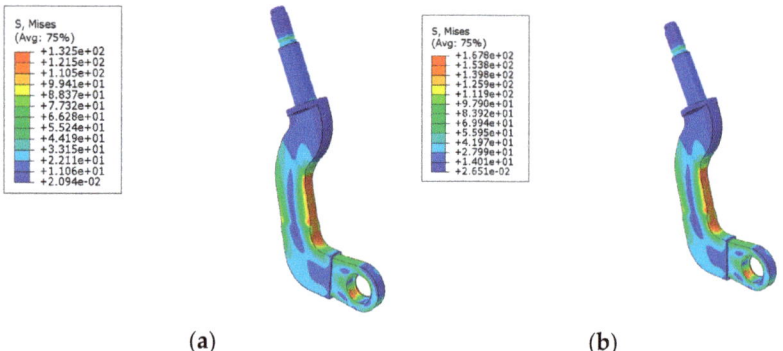

(a) (b)

Figure 10. Stress distribution in the boom under two operating conditions. (**a**) Stress nephogram of the boom at rated working conditions; (**b**) Stress nephogram of the boom at start-up.

As can be seen from the calculation results, in the rated and start-up two calculation conditions: boom maximum stress value is: 132.5 MPa, 167.8 MPa, maximum stress values are less than the boom material yield strength (≥490 MPa).

4.3. Fatigue Strength Assessment Methods

According to the relevant standards UIC 615-4 and TB/T 3548-2019, the Goodman diagram method is mainly used for fatigue strength assessment [21,22]. Fatigue strength assessment: select each node on the boom finite element model, simplify the stress state at each point into a uniaxial stress state based on the direction of the maximum principal stress for each working condition, calculate the stress value at each point σ_{max} and the minimum σ_{min} and calculate the equivalent average stress and equivalent force amplitude at each point according to the following formula:

$$\sigma_m = \frac{\sigma_{max} + \sigma_{min}}{2} \tag{1}$$

$$\sigma_a = \frac{\sigma_{max} - \sigma_{min}}{2} \tag{2}$$

For each working condition, the relatively dangerous nodes on the boom are selected, and the maximum and minimum stress values of these nodes are calculated under the positive and negative rotation of the gearbox and different vibration loads. The average stress and stress amplitude of each point is calculated according to the above method, and the equivalent average stress and equivalent stress amplitude of each node are put into the Goodman diagram [23] for fatigue strength assessment. The fatigue strength assessment results of each node are shown in Figure 11, which shows that the nodes selected on the boom fall within the Goodman fatigue limit, and the fatigue strength of the boom meets the design requirements.

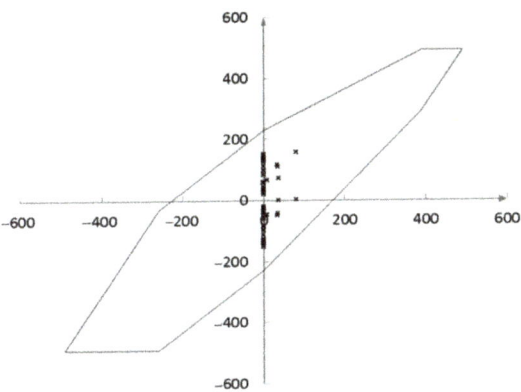

Figure 11. Fatigue strength assessment of booms.

5. Quantitative Fracture Analysis

The technique of studying the fracture surface of metal components is also one of the basic tasks and important methods of failure analysis [24]. The quantitative analysis of the fracture surface is used to determine the fatigue crack propagation rate of the component in actual operation, to provide a reasonable estimate of the life of fatigue crack growth in the component and to provide a reference for determining the cause of component failure [25].

The force analysis of the boom is mainly subjected to torque and vibration stresses. As the metro starts to stop (each stop), the torque increases first, smoothly in the middle and then decreases in a changing pattern, and the torque state of the metro running at each stop can be equated to a trapezoidal wave. The vibration stress during boom operation is a random fluctuation and can be equated to a triangular wave. Therefore, the stress variation curve of the boom can be equated to a trapezoidal wave superimposed on a triangular wave, see Figure 12.

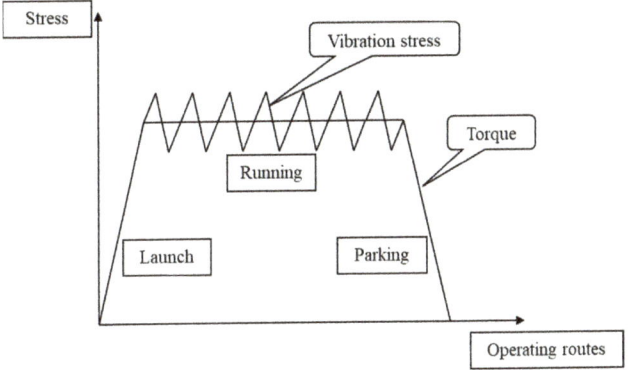

Figure 12. Boom force analysis per station.

Combined with the observation results of the boom fracture, the boom fracture shows the shape of fatigue beach marks and fatigue striations, which correspond to the torque and vibration stresses that the boom is subjected to. Therefore, the fatigue beach marks on the fracture are used to quantitatively analyze the life of fatigue crack growth in the boom [26].

The fatigue beach marks on the fracture were observed and analyzed, and the statistical results are shown in Table 4. A total of 819 fatigue beach marks were analyzed using the mean value method for the life of fatigue crack growth in the boom, and the number of fatigue beach marks in the extended phase of the boom was 819.

Table 4. Data relating to fatigue beach mark spacing at the boom break.

No.	Length from the Origin of the Fracture a_n (mm)	Average Distance between Beach Marks (mm)	N_i
1	0~24	0.10	240
2	24~47	0.10	230
3	47~82	0.12	292
4	82~90	0.14	57
5	-	-	$\Sigma N_n = 819$

The metro line has a total of 18 stations and a route length of 26.6 km, with an average distance of approximately 1.5 km per station. Quantitative analysis of fractures gives 819 fatigue beach marks for failed boom fractures, with a life of fatigue crack growth of 819 stations running, corresponding to a running distance of 819 stations ÷ 18 stations × 26.6 km = 1210 km.

6. Discussion

Through the above analysis, it can be seen that:

(1) The boom fractures at the transition of the structural bend, the origin of the fracture, are located at the inner edge of the bend, and typical fatigue beach marks and fatigue striations are visible in the extended area. From this, the nature of the fracture of the boom can be judged as fatigue.

(2) The boom originates at the inner edge of the bend, indicating a relatively high initial stress on the boom. A large number of rapid tearing features are also visible in the middle and late stages of the spreader bar fracture extension, also indicating high stress on the spreader bar. According to the commissioner, the metro line has high motor power and is subject to higher forces, approximately 10% to 20% higher than other lines, and three same-mode failures occurred between 360,000 km and 560,000 km of operation (the line runs a total of 39 trains), which was analyzed as possibly being a low boom fatigue tolerance in the line operating condition. From the metallographic analysis, it can be seen that there is a decarburization layer in the range of 0.3 mm on the surface of the boom, with a hardness of 167 HB in the decarburization zone, much lower than the hardness of the core (230 HB), which plays a role in promoting fatigue cracking of the boom.

(3) By finite element calculation, the maximum stress point of the boom under the rated and start-up conditions is located at the fracture corner. The maximum stress values of the boom under the rated and start-up conditions are 132.5 MPa and 167.8 MPa respectively, and the maximum stress values are less than the yield strength of the boom material ($\geq$490 MPa).

(4) The results of the quantitative analysis of the fracture indicate that the life of fatigue crack growth in the boom is 819 stations (equivalent to 1210 km), according to the fatigue initiation life is equal to the total life minus the life of fatigue crack growth, it is known that the fatigue initiation life of the boom is (523,662 − 1210) km = 522,452 km.

7. Conclusions

The fatigue life margin of the boom is low, which is the main reason for fatigue fracture. The decarburization of the shallow surface also promotes the fatigue fracture of the suspension rod in the metro gearbox.

The recurrence of this boom fracture problem can be avoided from the following aspects.

In order to prevent decarburization and not affect the performance of the boom, the machining allowance at the corner of the boom is increased so that the depth of the decarburized layer is less than the machining allowance and can be completely cut off when mechanical machining is carried out. The correct heat treatment process operation can then be strictly implemented.

In order to reduce the magnetic particle testing cycle of the boom in service, the boom with fatigue cracks was replaced in time and all the booms of the batch were replaced during the subsequent overhaul period.

Author Contributions: Z.X. conducted the research and analyzed the results under the supervision of X.L. and W.L. Z.X. contributed to the design of experiments and morphology analysis during the research activities. Z.X. and H.L. contributed to image processing and data calculation. All authors have read and agreed to the published version of the manuscript.

Funding: The authors are grateful for the support provided by the National Natural Science Foundation of China, grant number 52075050, Natural Science Foundation of Jiangsu Province, grant number BK20201448, Natural Science Foundation of Jiangsu Province, grant number BK20210854, Natural Science Fund Project of Colleges in Jiangsu Province, grant number 20KJB470009 and Postgraduate Research & Practice Innovation Program of Jiangsu Province, grant number KYCX22_3022.

Institutional Review Board Statement: Not applicable.

Informed Consent Statement: Not applicable.

Data Availability Statement: Not applicable.

Conflicts of Interest: The authors declare no conflict of interest.

References

1. Ruobing, K.; Zhigang, K. Structural Strength Analysis of High Current Connector for Rail Transit. *J. Phys. Conf. Ser.* **2021**, *2095*, 012013.
2. Mejuto Diego, G. Theorizing nation-building through high-speed rail development: Hegemony and space in the Basque Country, Spain. *Environ. Plan. A Econ. Space* **2022**, *54*, 554–571. [CrossRef]
3. Wangang, Z.; Wei, S.; Hao, W. Vibration and Stress Response of High-Speed Train Gearboxes under Different Excitations. *Appl. Sci.* **2022**, *12*, 712. [CrossRef]
4. Li, X.; Yang, Y.; Wu, Z.; Yan, K.; Shao, H.; Cheng, J. High-accuracy gearbox health state recognition based on graph sparse random vector functional link network. *Reliab. Eng. Syst. Saf.* **2022**, *218*, 108187. [CrossRef]
5. Feng, Y.; Xiaochun, Z.; Chunyu, L.; Jun, L. Analysis and Research on Fracture Cause of Fixed Shaft of Torsion Arm of Wind Turbine Gearbox. *J. Phys. Conf. Ser.* **2021**, *2133*, 012039. [CrossRef]
6. Hu, Z.; Yang, J.; Yao, D.; Wang, J.; Bai, Y. Subway Gearbox Fault Diagnosis Algorithm Based on Adaptive Spline Impact Suppression. *Entropy* **2021**, *23*, 660. [CrossRef] [PubMed]
7. Curà, F.; Mura, A.; Rosso, C. Effect of rim and web interaction on crack propagation paths in gears by means of XFEM technique. *Fatigue Fract. Eng. Mater. Struct.* **2015**, *10*, 1237–1245. [CrossRef]
8. Meneghetti, G.; Dengo, C.; Lo Conte, F. Bending fatigue design of case-hardened gears based on test specimens. *Proc. Inst. Mech. Eng. Part C J. Mech. Eng. Sci.* **2018**, *11*, 1953–1969. [CrossRef]
9. Conrado, E.; Gorla, C.; Davoli, P.; Boniardi, M. A comparison of bending fatigue strength of carburized and nitrided gears for industrial applications. *Eng. Fail. Anal.* **2017**, *78*, 41–54. [CrossRef]
10. Shukla, A.; Rai, P. Finite element based modelling and analysis of dragline boom structure. *J. Mines Met. Fuels* **2020**, *68*, 50–56.
11. Zalaznik, A.; Nagode, M. Experimental, theoretical and numerical fatigue damage estimation using a temperature modified dirlik method. *Eng. Struct.* **2015**, *96*, 56–65. [CrossRef]
12. Irisarri, A.M.; Pelayo, A. Failure analysis of an open die forging drop hammer. *Eng. Fail. Anal.* **2009**, *16*, 1727–1733. [CrossRef]
13. Lu, Y.; Ripplinger, K.; Huang, X.J.; Mao, Y.; Detwiler, D. A new fatigue life model for thermally-induced cracking in H13 steel dies for die casting. *J. Mater. Process. Technol.* **2019**, *271*, 444–454. [CrossRef]

14. Mellouli, D.; Haddar, N.; Köster, A.; Ayedi, H.F. Thermal fatigue failure of brass die-casting dies. *Eng. Fail. Anal.* **2012**, *20*, 137–146. [CrossRef]
15. Cui, Z.; Bhattacharya, S.; Green, D.E.; Alpas, A.T. Mechanisms of die wear and wear-induced damage at the trimmed edge of high strength steel sheets. *Wear* **2019**, *427*, 1635–1645. [CrossRef]
16. General Administration of Quality Supervision; Inspection and Quarantine of the People's Republic of China. *GB/T 10561-2005 Determination of the Content of Non-Metallic Inclusions in Steel Standard Rating Chart Microscopic Test Method*, 1st ed.; China Standard Publishing House: Beijing, China, 2015.
17. China Iron and Steel Industry Association. *GB/T 228.1-2021 Metallic Materials—Tensile Testing—Part 1: Method of Test at Room Temperature*, 1st ed.; China Standard Publishing House: Beijing, China, 2021.
18. China Iron and Steel Industry Association. *GB/T 231.1-2018 Metallic Materials—Brinell Hardness Test—Part 1: Test Method*, 1st ed.; China Standard Publishing House: Beijing, China, 2018.
19. China Iron and Steel Industry Association. *GB/T 229-2020 Metallic Materials—Charpy Pendulum Impact Test Method*, 1st ed.; China Standard Publishing House: Beijing, China, 2020.
20. Statharas, D.; Sideris, J.; Chicinas, I.; Medrea, C. Microscopic examination of the fracture surfaces of a cold working die due to premature failure. *Eng. Fail. Anal.* **2011**, *18*, 759–765. [CrossRef]
21. International Union of Railways. *UIC 615-4 Motive Power Units-Bogies and Running GearBoige Frame Structure Strength Tests*, 1st ed.; International Union of Railways: Paris, France, 2003.
22. China Railway Bureau. *TB/T 3548-2019 General Rules for Strength Design and Test Qualification Specifications for Rolling Stock*, 1st ed.; China Standards Publishing House: Beijing, China, 2019.
23. Institute of Metals and Chemistry; Ministry of Railways Scientific Research Institute. *The p-s-N Curves and Goodman's Atlas for Materials Commonly Used in Railways*, 1st ed.; Research Report of the Academy of Railway Sciences: Beijing, China, 1999.
24. Liao, Z.; Wenfeng, Z.; Hai, Y. Application of quantitative fracture analysis in assessing the fatigue life of components. *Mater. Eng.* **2000**, *4*, 45–48.
25. Xinling, L.; Weifang, Z.; Chunhu, T. Comparison of fatigue life models for quantitative analysis of different fractures. *Mech. Eng. Mater.* **2008**, *5*, 4–6.
26. Xin, W.; Chunling, X.; Xing, C.; Dianyin, H.; Bo, H.; Rengao, H.; Yuanxing, G.; Zhihui, T. Effect of cold expansion on high-temperature low-cycle fatigue performance of the nickel-based superalloy hole structure. *Int. J. Fatigue* **2021**, *151*, 106377.

Article

Effect of Cladding Current on Microstructure and Wear Resistance of High-Entropy Powder-Cored Wire Coating

Xinghai Shan [1], Mengqi Cong [2,*] and Weining Lei [1]

[1] School of Mechanical Engineering, Jiangsu University of Technology, Changzhou 213001, China
[2] Key Laboratory of Advanced Materials Design and Additive Manufacturing of Jiangsu Province, Jiangsu University of Technology, Changzhou 213001, China
* Correspondence: congmq@jsut.edu.cn

Abstract: This paper investigated the effect of tungsten arc melting current on the microstructure and wear resistance of coatings prepared from high-entropy powder-cored wire, FeCrMnCuNiSi$_1$. A powder-cored wire of high-entropy composition was drawn by powder-cored wire-forming equipment, and a FeCrMnCuNiSi$_1$ high-entropy alloy coating was designed on the base material 40Cr by the tungsten arc fusion technique. The influence law and mechanism of melting current on the wear resistance of the coatings were obtained through analyzing the microstructure, physical phase, and wear resistance of the coatings prepared by different melting currents. At a melting current of 200A, the FeCrMnCuNiSi$_1$ coating exhibits fine equiaxed grains and a single BCC phase; the highest and average microhardness of the coating reach 790.36 HV and 689.73 HV, respectively, whose average microhardness is twice that of the base material. The wear rate of the coating is 2245.86 μm^3/(N·μm), which is only 8% of the base material and has excellent wear resistance. The FeCrMnCuNiSi$_1$ high-entropy alloy coating prepared by ordinary powder-cored wire-forming equipment and the tungsten arc cladding method has excellent performance and low cost, which can provide an essential basis for the development, preparation, and application of high-entropy alloy coatings.

Keywords: coating; high-entropy alloy; tungsten arc; fatigue wear; abrasive wear; adhesive wear

1. Introduction

While conventional alloys have been endowed with various desirable properties, usually by adding relatively few minor elements to the major ones, high-entropy alloys break with this concept and consist of high concentrations of multiple major elements [1]. In 2004, Yeh et al. [2] and Cantor et al. [3] introduced the concepts of high-entropy alloys and isotonic ratio multicomponent alloys, respectively, which are alloyed materials composed of five or more significant elements with atomic fractions ranging from >5% to <35% of each central element. The high-entropy alloy thermodynamically exhibits a lower Gibbs free energy because of its mixed state with multiple primary elements. The microstructure of high-entropy alloys mainly consists of face-centered cubic (FCC), body-centered cubic (BCC), HCP (dense hexagonal), and a few compounds generated from mixed elements. The alloys have some excellent properties that are incomparable to traditional alloys, such as high strength [4], high wear resistance [5], corrosion resistance [6], high tensile strength [7], and resistance to oxidation at high temperatures [8]. Therefore, high-entropy alloys have become hot spots for research as coatings applied to the surface modifications of steel, aluminum alloys, and other materials.

At present, the preparation methods of high-entropy alloys mainly include vacuum arc melting [9], induction arc melting [10], mechanical alloying [11], and magnetron sputtering [12]. However, these methods have the problems of a complex production process and high preparation costs. Therefore, the coating preparation method with simple operation

and low manufacturing cost has attracted the attention of production practices, such as laser cladding [13], electron beam melting [14], plasma arc deposition [15], and tungsten arc cladding [16]. For example, Zoia Duriagina et al. used laser alloying and plasma chemical vapor deposition to prepare coatings with excellent performance on stainless steel [17,18]. In these coating preparation methods, tungsten arc cladding has the characteristics of low equipment cost, easy operation, and comprehensive application scenarios. In addition, due to the cladding process, the tungsten needle does not melt and the discharge arc length changes relatively few disturbing factors, as well as the welding process is stable and the whole process is filled with the inert shielding gas argon, which forms a good airflow isolation layer that effectively prevents oxygen and nitrogen from equaling the coating development into a chemical reaction; thus, the prepared coating has better fusion with the substrate, which has received much attention from researchers. Shen et al. [19] prepared cable-type filaments. They used a tungsten arc to prepare a bulk AlCoCrFeNi high-entropy alloy with no defects in the microstructure of FCC and BCC; its compressive strength reached 2.9 GPa, and its elongation reached 42%. Dong et al. [20] prepared their powder bed on an arc cladding platform to place the proportioned high-entropy alloy powder, which prevents the protective gas from blowing away the preplaced powder. The high-entropy alloy AlCoCrFeNi$_{2.1}$ prepared by this device has a good combination of tensile strength (719 MPa) and flexibility (27%), with a pseudo-eutectic microstructure consisting of micro phases of more significant FCC phase columnar grains (90 wt%) and a refined BCC phase (10 wt%). Fan et al. [21] investigated the preparation of high-entropy alloy coatings by ultrasound-assisted tungsten arc melting. The average grain diameter of the coatings was reduced from 285 μm to 78 μm, which was refined by 70% under the ultrasonic treatment. The microhardness was increased by 20%, which increased from 441 HV to 532 HV. The optimization of the properties of the high-entropy alloy originated from the grain refinement after the cavitation effect of the ultrasonic treatment. When a high-entropy alloy coating is prepared by arc fusion coating, there are still problems, such as the large diameter of the stranded wire, which cannot be satisfied by existing automatic wire-feeding equipment, significant burn loss during the prefabricated powder fusion coating process, and poor coating formability.

To avoid the above defects in preparing high-entropy alloy coatings by the arc melting method, a new high-entropy powder-cored wire material, FeCrMnCuNiSi$_1$, was developed independently. A commonly used component configuration, 40Cr steel, which has good mechanical properties and welding fusion performance, is used for the base material. The high-entropy alloy coating was prepared using a tungsten arc platform and an automatic wire-feeding device. This method effectively avoids the preset powder being blown away by the protective gas during the melting process. In addition, the automatic wire-feeding device can accurately feed the wire to prepare a high-entropy alloy coating with a controlled thickness. At the same time, the microstructure and mechanical properties of the high-entropy alloy coatings were further investigated, which could provide new ideas and ways to prepare a new high-entropy alloy coating.

2. Materials and Methods

Figure 1 shows a schematic diagram of the process of the FeCrMnCuNiSi$_1$ high-entropy alloy coating. Firstly, according to the design concept of the high-entropy alloy isotonic ratio, the configured alloy powder is placed into a ball mill tank with an appropriate ball material ratio for uniform mixing, and the ball mill tank is filled with the protective gas argon to prevent a chemical reaction of the alloy powder during the ball milling process. Then, the mechanically mixed alloy powder is filtered and dried, followed by placing it in the hopper of the powder-cored wire-forming equipment. A 304 stainless steel strip is rolled into a U-shaped open tube by the forming roller on the powder-cored wire-forming equipment. Finally, the U-shaped open tube is covered with mixed powder, and the welding wire with the specified diameter is obtained through cold rolling, tube rolling, and wire drawing. A 40Cr is used as the substrate material, with a size of 200 mm × 200 mm ×

10 mm. Before melting, the surface of the substrate should be sanded clean with 400- and 800-purpose sandpaper to remove any oil and rust, which can improve the fusion effect of the coating and the substrate. Meanwhile, the platform surface of the arc table should be flat to ensure good conductivity. The optimal high-entropy alloy coatings prepared by the powder-cored wire FeCrMnCuNiSi$_1$ were explored by adjusting the parameters of the melting current. The parameters of the arc cladding were as follows: the shielding gas argon was 15 L/min, the working voltage was 10 V, the arc length was 4 mm, the automatic wire-feeding speed was 150 mm/min, and the cladding currents were 200 A, 220 A, and 240 A, respectively.

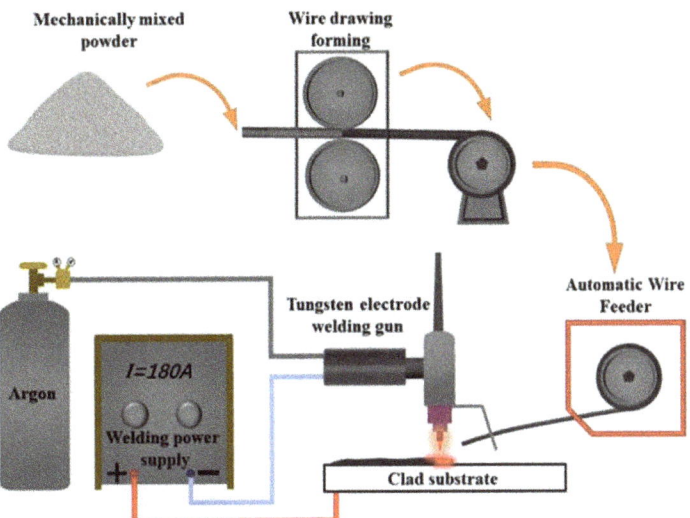

Figure 1. Schematic diagram of the process of the FeCrMnCuNiSi$_1$ high-entropy alloy coating.

The specimens with sizes of 10 mm × 10 mm and 20 mm × 20 mm were cut from the molten coating by an SG3000 EDM machine for surface microstructure characterization and mechanical property testing. The cutting specimen was polished and cleaned by an ultrasonic cleaner, followed by being corroded with a corrosive acid (HNO$_3$:HCl = 1:3) for 10 s. The above operation was repeated for the specimens with a size of 20 mm × 20 mm, but without a corrosion treatment.

A metallurgical microscope and a SIGMA500 type field-emission scanning electron microscope were used for observing the microstructure before and after the corrosion treatment. The phase analysis of the coatings was analyzed by an HD-Xpretty PRO type X-ray diffractometer with operating parameters of 40 KV and 20 mA and a scanning speed of 5°/min. An HVS-1000A hardness tester was used to test the microhardness of the coating with an applied load of 200 gf and a holding time of 15 s. The wear resistance of the coating was tested for 10 min by an MDW-02 high-speed reciprocating friction and a wear tester with a load of 50 N and a frequency of 10 Hz. A Si$_3$N$_4$ with a diameter of 6 mm was used as the grinding ball. The wear volume was calculated. Furthermore, the wear surface morphology and profile were further observed by a Naonvea PS50 3D profiler to investigate the wear mechanism of the coating.

3. Results and Discussion

3.1. Microstructure Characterization

Figure 2 shows the optical micrographs of the FeCrMnCuNiSi$_1$ coating by tungsten arc melting at different currents. Figure 2a–c show the morphology of the coating at three current parameters (200 A, 220 A, and 240 A), all of which exhibit good cross-sectional profiles with no pores or cracks. Different microstructure characteristics of the region [A] near the fusion line can be observed cleanly in Figure 2(a_1,b_1,c_1). The microstructure of the fusion line between the coating and the substrate exhibits a columnar grain morphology, and the columnar crystals extend to the substrate. Because the substrate subjacent to the melt pool conducts heat as the heat dissipates during the melting, a large number of heterogeneous nucleation are formed under the temperature gradient. Cubic crystals with the orientation of <001> produce epitaxial solidification and fast growth, while other grain growth is hindered. Thus, the coarse columnar grains of the region [A] near the fusion line gradually become larger with the increasing current, which is consistent with the grain forming characteristics of the arc fusion zone and coating [22].

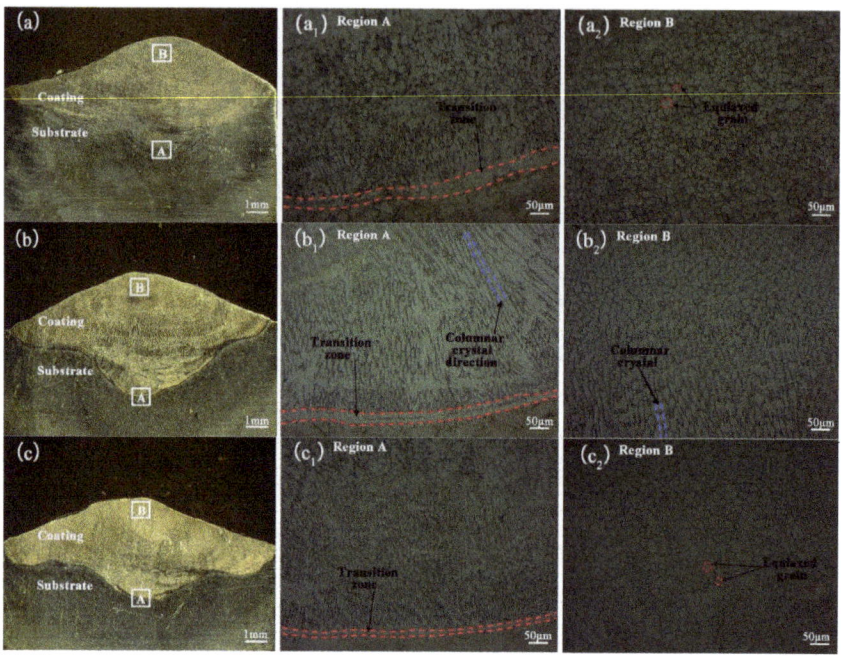

Figure 2. Optical micrographs of the coatings on the FeCrMnCuNiSi$_1$ powder-cored wire at different cladding currents. (**a**,**a_1**,**a_2**) 200 A; (**b**,**b_1**,**b_2**) 220 A; (**c**,**c_1**,**c_2**) 240 A.

The top region [B] of the coating in Figure 2a and c is a more uniform equiaxed crystal structure, while the top region [B] in Figure 2b is a columnar crystal growing by means of dendritic growth. The top of the coating shows a mixture of refined equiaxed and columnar crystals at a melting current of 240 A. The increase in melting current at a constant melting speed leads to a widening at the bottom of the melt pool and an increase in the contact area between the melt pool and the substrate, which increases the ability of the dissipated heat from the melt pool to the substrate and the gradual refinement of grains at the fusion line [23].

The microstructure of the medium-axis and columnar crystals at different melting currents and the typical dendrite microstructure (DR) and interdendritic microstructure (ID) were further observed through scanning electron microscopy, as shown in Figure 3. According to the EDS analysis of the elemental content of the dendritic and interdendritic tissues inside the coating obtained in Table 1, it was found that the elemental content of Fe in the interdendritic tissues is higher than that in the dendritic tissues. The Fe element does not affect the solid solution phase and microstructure, which is significantly higher than the nominal value of the high-entropy alloy due to the dilution effect of the base material.

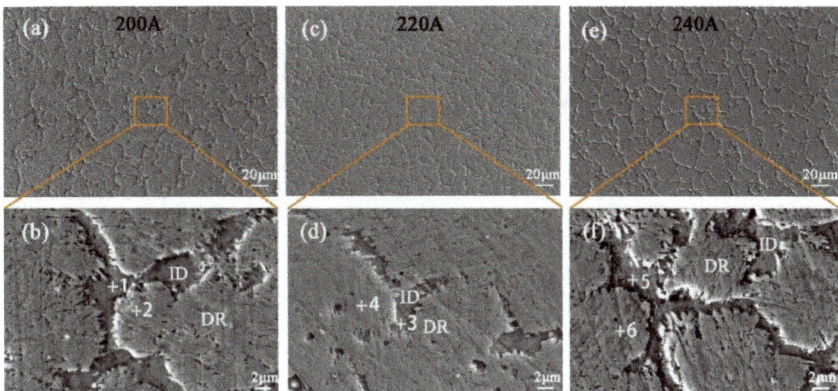

Figure 3. (**a,c,e**) show the SEM of the top of FeCrMnCuNiSi$_1$ high-entropy alloy coating at melting currents of 200 A, 220 A, and 240 A, respectively. (**b,d,f**) show the enlarged views of the orange rectangular areas in (**a,c,e**).

Table 1. EDS of the microstructure of the top region of the FeCrMnCuNiSi$_1$ high-entropy alloy coating at different melting currents.

Coating	Point	Fe	Cr	Mn	Cu	Ni	Si
200 A	1	41.8	17.4	13.9	10.2	13.5	1.1
	2	54.2	13.7	11.6	8.4	10.2	0.6
220 A	3	63.6	9.8	8.8	5.6	8.4	1.9
	4	68.7	7.8	8.2	4.3	8.1	1.4
240 A	5	52.9	15.2	13.5	7.5	8.6	1.0
	6	62.5	10.1	10.0	6.7	8.3	0.9

The Cr elements are mainly in the interdigitated dendritic region, and the ease of the Cr element deviation may be due to its relatively low migration activation energy [24–26]. At the same time, the Cr element can promote the generation of a BCC solid solution, and the hardness and strength of the coating may be reduced due to the decrease in the Cr element content by increasing the current. The relative content of the Cu element is low at 220 A and 240 A. Combined with the analysis in Figure 3b, the melt pool temperature is high, and the amount of Cu element burned is severe due to the higher current, the slower heat dissipation, and the smaller contact area between the melt pool and the base material. It can be observed from Figure 3d that there is a small area of interdendritic tissue. The EDS results in Table 1 show that the elemental content of the dendritic tissue and interdendritic tissue of the columnar crystal is relatively uniform and the elemental segregation is relatively tiny.

3.2. Phase Constitution

Figure 4 shows the XRD patterns of the FeCrMnCuNiSi$_1$ high-entropy alloy coatings with different currents. The XRD patterns show that the FeCrMnCuNiSi$_1$ coating exhibits a BCC phase at 44.76°, 65.15°, and 82.52° with a lattice constant of 2.868 Å, which is basically the same as the α-Fe or CrFe$_4$ lattice constants (λ = 2.866 Å, PDF#87-0721; λ = 2.866 Å, PDF#65-7251). The α-Fe or CrFe$_4$ phases are formed by the A2 (W-type, disordered BCC) structure and B2 (β'-CuZn-type, ordered BCC) structure, respectively [27]. For the FCC diffraction peak at 51.15°, the dot constant of 10.093 Å is the same as that of Fe$_5$Ni$_4$S$_8$ (λ = 10.093 Å, PDF#86-2470). The diffraction peak of the coating at a melting current of 200 A exhibits a single BCC structure. The coating prepared at 220 A exhibits the FCC phase. It can also be found that the angle of the FCC phase diffraction peak gradually widens with the increase in the cladding current, which is due to the rise in the cladding current and the increase in the melt pool temperature. The Ni element, which has a high melting point in the powder-cored wire, reacts with the small amount of Si element in the base material to produce $Fe_5Ni_4S_8$ with a high melting point. The grain size of the coating was calculated by the Debye–Scherrer formula. With the increase in the current from 200 A to 240 A, the grain sizes at the orientation (110) were 15.31, 15.52, and 16.28 nm, respectively, and the grain sizes at the direction (400) were 0, 15.31, and 16.21 nm, respectively. Thus, the grain size of the coating increases with the increase in the current, which verifies the phenomenon that the grain shape becomes more significant with the rise in current in the optical micrograph.

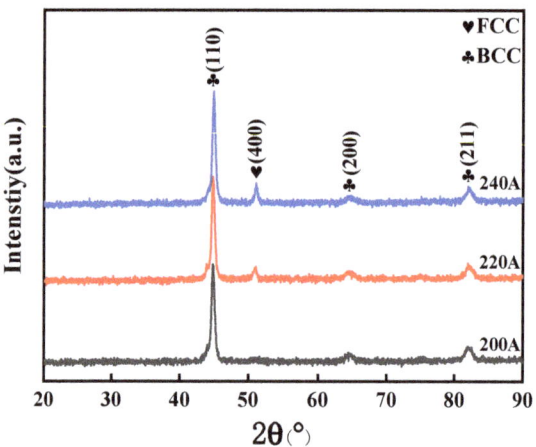

Figure 4. XRD patterns of the $FeCrMnCuNiSi_1$ high-entropy alloy coatings with different currents.

3.3. Mechanical Properties

3.3.1. Microhardness

Figure 5 shows the cross-sectional microhardness results of the coatings at different melting currents. The histograms and line graphs show the average microhardness and microhardness values at each test point on the coating and the substrate, respectively. The histogram shows that the average microhardness of the base material 40Cr is 343.62 HV. The coating has the highest average microhardness of 689.73 HV at a melting current of 200 A, which is twice the average hardness of the base material. It is obvious from the line graph that the microhardness of the coating with the increasing current has a decreasing trend. Because the internal temperature of the melt pool increases sharply, the contact area between the melt pool and the substrate decreases, and the cooling rate of the coating decreases sharply with the increasing current, which results in a larger grain size of the coating. According to the Hall–Patch formula, the grain size increases and the strength and hardness decrease [28]. Additionally, the coatings at the melting currents of 220 A and 240 A have a lower content of non-metallic Si elements, which will improve the plasticization of the coatings and reduce their hardness. The coatings at these two currents derive from FCC phases and have a lower content of non-metallic Si elements; FCC phases and Si elements in high-entropy alloys can improve the plasticization; thus, the hardness decreases due to the combined effect of grain size, FCC, and Si elements. The coating has the lowest average microhardness when the melting current is 220 A. The coating at the melting current of 220 A exhibits a columnar crystal structure, which has a lower Cr segregation compared to the equiaxed crystals, resulting in the lowest strength and hardness. It can also be observed that the maximum hardness of the coatings appears at the top of the coating, which is due to the diffusion strengthening effect through the refinement of the grain microstructure in the top region. At the same time, the fusion lines are mostly coarse columnar crystals extending toward the base material, which leads to the lowest hardness [29].

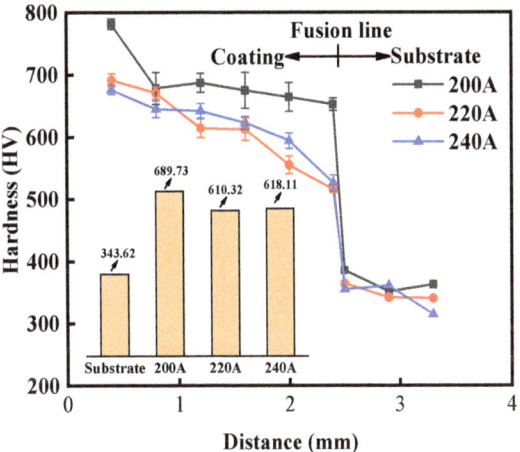

Figure 5. Cross-sectional microhardness of the FeCrMnCuNiSi$_1$ high-entropy alloy coatings with different currents.

3.3.2. Abrasion Resistance

Figure 6 shows the 3D profiles of the wear tracks for the base material and the high-entropy alloy coating at different melting currents. It can be observed that the wear track in the base material is much wider than that of the high-entropy alloy coating. Combined with the color scale on the right side in Figure 6, it is found that the wear track of the base material is much deeper than that of the coating. The data in Table 2 can be obtained by analyzing the 3D profiles. The wear tracks of the coatings reach a wear depth of 11.6, 18.4, and 15.5 μm at the melting current of 200 A, 220 A, and 240 A, respectively. The cross-sectional areas of the coating wear tracks were 4133, 7791, and 4052 μm^2, respectively. The wear was most serious at the melting current of 220 A, but its depth was only 1/5 of the cross-sectional area of the base material. The wear resistance of the coating can be compared quantitatively by the wear rate, which Equation (1) shows as follows:

$$\delta = V/\Sigma W, \tag{1}$$

$$Q = KWL/H, \tag{2}$$

(1) where V is the wear volume, and ΣW is the accumulated work done by friction;
(2) where Q is the material wear volume, K is the friction coefficient, W is the normal load, L is the sliding length, and H is the material surface hardness.

Table 2. Wear resistance test results of the base material and the high-entropy alloy coating of FeCrMnCuNiSi$_1$ with different melting currents.

Sample	Maximum Wear Depth μm	Wear Cross-Sectional Area μm^2	Wear Rate μm^3/(N·μm)
40Cr	52.5	39,765	26,984.86
200 A	11.6	4133	2999.76
220 A	18.4	7791	6168.82
240 A	15.5	4052	2245.86

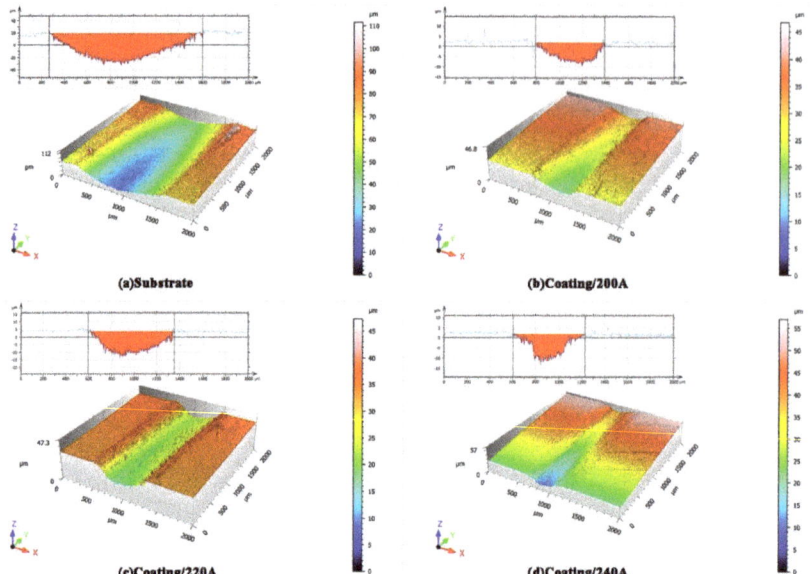

Figure 6. (**a**) 3D view and cross-sectional view of the abrasion marks of the base material. (**b**–**d**) 3D view and cross-sectional view of the abrasion marks of the FeCrMnCuNiSi$_1$ high-entropy alloy coating with different melting currents.

The wear rate, which is obtained from Equations (1) and (2), at the melting current of 200 A, 220 A, and 240 A is 2999.76 µm^3/(N·µm), 6168.82 µm^3/(N·µm), and 2245.86 µm^3/(N·µm), respectively. It can be observed that the maximum wear rate is only 22% of the base material wear rate. From the above analysis, the coating has the lowest average microhardness when the melting current is 220 A. According to the calculation formulas [30], the amount of material wear is inversely proportional to the hardness of the material under the same external conditions. thus, the coating at the melting current of 220 A has a larger amount of wear.

To further investigate the effect of the melting current on the wear resistance of the coating, the wear morphologies of the substrate and the FeCrMnCuNiSi$_1$ high-entropy alloy coatings were observed and analyzed. Figure 7 shows the SEM images of the wear morphologies of the base material and the high-entropy alloy coatings at different melting currents. A large area of dark-colored flakes within the substrate abrasion marks can be found in Figure 7a. A further magnified Figure 7b shows that the abrasion flakes are nearly fan-shaped or flaky, and exhibit warped boundaries, which lead to the large craters after the abrasion flakes. This is because the coating surface is subjected to cyclic contact stress; the maximum shear stress is at a certain depth under the surface. When the surface strength is insufficient, fatigue cracks will be generated, at which time the maximum shear stress will extend to the surface along the plastic deformation, resulting in spalling of the surface material, which is characteristic of fatigue wear. At the same time, the contact surface temperature is high, and it is easy to produce an oxide layer because of the cyclic effect of stress, which can be proved by the EDS analysis of point 1 in Figure 7c. The oxygen content of point 1 reaches 15.5%, which is shown in Table 3. It can be concluded that the wear mechanism is typical of an oxide layer.

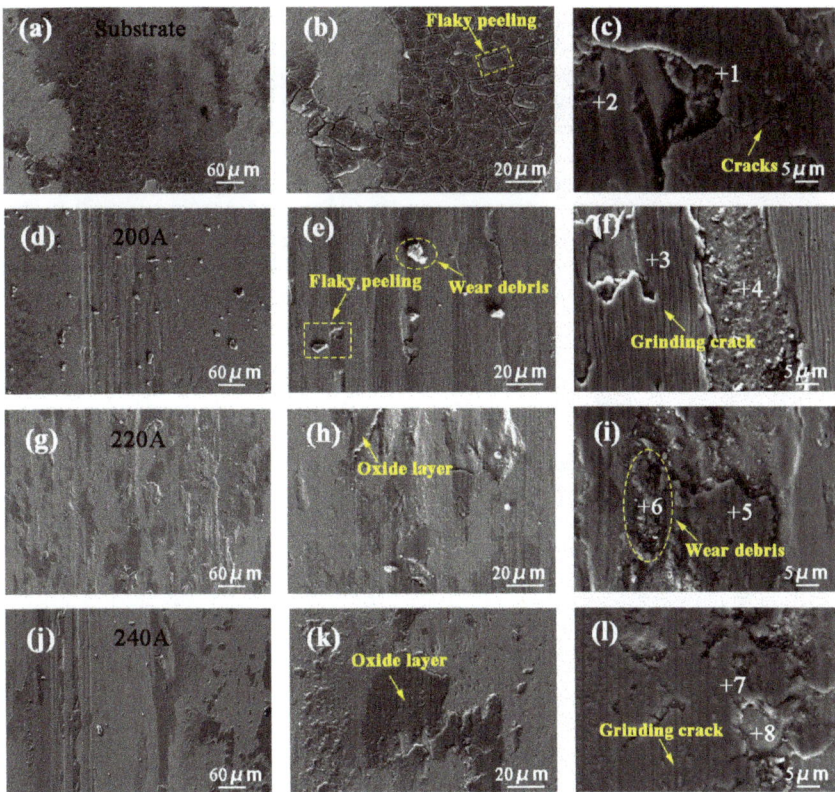

Figure 7. SEM images of the wear morphology of the 40Cr and FeCrMnCuNiSi$_1$ high-entropy alloy coatings at different melting currents. (**a–c**) Substrate; (**d–f**) 200 A; (**g–i**) 220 A; (**j–l**) 240 A.

Table 3. EDS of the friction and wear scar structures of the FeCrMnCuNiSi$_1$ high-entropy alloy coating under the substrate and different cladding currents.

Sample	Point	Fe	Cr	Mn	Cu	Ni	Si
40Cr	1	80.3	15.5	1.4	1.4	0.9	0.2
	2	96.2	0	1.9	1.0	0.7	0
200A	3	60.3	7.9	0.9	7.6	8.1	6.5
	4	67.1	0.6	1.3	7.4	8.5	6.3
220A	5	57.2	17.1	1.1	5.8	7.7	4.2
	6	66.2	1.9	1.2	5.4	8.1	7.4
240A	7	53.6	13.9	1.1	9.0	8.2	5.8
	8	66.3	0	1.2	8.6	7.3	5.4

It can be seen from Figure 7g that there is a relatively flat and smooth adhesion layer on the wear track of the coating at 220 A. It is in the same direction as the reciprocal sliding of the grinding ball, from which it can be concluded that the adhesion layer is a layer produced by the plastic deformation of the coating surface material caused by the applied load and shear stress. There are pits on the surface of the coating, as shown in Figure 7h. This may be caused by cracks in the stress concentration area of the surface layer of the coating under the cycling action of Si$_3$N$_4$, after which the cracks can extend to the surface and peel off to form pits. An EDS analysis of points 5 and 6 in Figure 7i, shown in Table 3, revealed that the oxygen content at point 5 was significantly higher than that of point 6, which reached 17.1%; thus, the dark adherent layer on the wear path of the coating at

220 A was judged to be an oxide layer. From the above, it is known that the coating has a high Si content at the melting current of 220 A. An oxide layer with a high Si content is easy to break and adhere to the surface of the Si_3N_4 grinding balls by an atomic bonding reaction under local stress (cold welding), resulting in a large amount of wear on the coating surface. Therefore, the wear mechanism of the coating at 220 A is adhesive wear. Uniformly scattered granular abrasive chips are found in the wear tracks of the coating at 200 A and 240 A, as shown in Figure 7d,j. More grooves of varying depths are also found on the wear surface, which were caused by the plowing action of the Si_3N_4 grinding balls. It is judged that these two coatings produce abrasive wear on the surface. The EDS results of Figure 7f,l are shown in Table 3. The oxygen content of the surface layer is much higher than that of the crater coating; thus, the surface layer undergoes oxidative wear. The shift in the wear mechanism of the coating exists because of the high microhardness of the coating at the melting currents of 200 A and 240 A. A non-material transfer of abrasive chips on the wear surface occurs, and the frictional heat promotes the oxidation of the microscopic abrasive chips, which eventually form an oxide layer under the continuous grinding of the Si_3N_4 grinding balls. Figure 7j,k show the presence of a larger oxide layer in the area on the wear surface at the melting current of 240 A.

Combined with the surface element distribution analysis in Figure 8, it can be observed that the oxide layer within the wear marks of this parameter is flat and dense, and the dense oxide layer in the high-entropy alloy coating is beneficial to the improvement of the wear resistance of the coating [31].

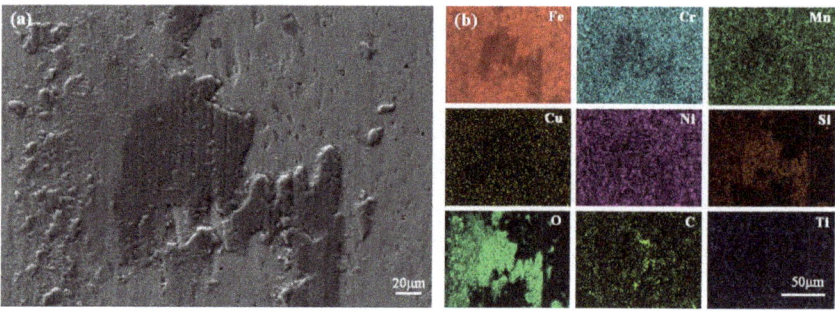

Figure 8. Wear morphology and surface element distribution of the high-entropy alloy coating at a melting current of 240 A. (**a**) Wear marks SEM; (**b**) Wear marks EDS surface scan.

4. Conclusions

By investigating the effect of the tungsten electrode melting current on the microstructure and wear resistance of the $FeCrMnCuNiSi_1$ coating, the process parameters of the coating prepared by the $FeCrMnCuNiSi_1$ high-entropy powder-core wire were obtained, and the microstructure, microhardness, and wear resistance of the coatings were studied. The main conclusions can be obtained as follows:

1. When the melting current is 200 A, the coating shows a single BCC structure. When the melting current is 220 A and 240 A, the coating is a mixed phase of FCC and BCC. When the melting current is 220 A, the coating microstructure is a coarse columnar crystal. When the melting current is 200 A and 240 A, the coating microstructure is a fine equiaxed crystal.
2. The coating has the highest average microhardness of 689.73 HV at a melting current of 200 A, which is twice the average hardness of the base material. As the current increases, the average microhardness of the coating decreases significantly with the decrease in Cr segregation in the coating and the generation of the Si-containing FCC phase.
3. The wear resistance of the coating is weakest at 220 A, which is only 1/3 of the maximum wear depth of the base material. The cross-sectional area of the abra-

sion marks is 1/5 of the cross-sectional area of the base material. The wear rate is 6168.82 $\mu m^3/(N \cdot \mu m)$, which is only 22% of the wear rate of the base material. The wear mechanism of the coating at 220 A is mainly adhesive wear and oxidation wear. When the melting current is 200 A, the prepared high-entropy alloy coatings have good wear resistance and meet the practical production requirements.

Author Contributions: Conceptualization, X.S. and M.C.; methodology, W.L.; formal analysis, X.S.; investigation, X.S.; resources, W.L.; writing—original draft preparation, X.S.; writing—review and editing, M.C. and W.L.; supervision, W.L.; project administration, W.L.; funding acquisition, W.L. All authors have read and agreed to the published version of the manuscript.

Funding: National Natural Science Foundation of China (Grant No. 51975264); Natural Science Foundation of the Jiangsu Higher Education Institutions of China (Grant No. 19KJD430004); Jiangsu University of Technology Graduate Student Practice Innovation Program (Grant No. XSJCX21_10).

Data Availability Statement: Not applicable.

Acknowledgments: The authors thank the Jiangsu Key Laboratory of Advanced Material Design and Additive Man-ufacturing for providing equipment support.

Conflicts of Interest: The authors declare that they have no known competing financial interest or personal relationships that could have appeared to influence the work reported in this paper.

References

1. George, E.P.; Raabe, D.; Ritchie, R.O. High-entropy alloys. *Nat. Rev. Mater.* **2019**, *4*, 515–534. [CrossRef]
2. Yeh, J.W.; Chen, S.K.; Lin, S.J.; Gan, J.Y.; Chin, T.S.; Shun, T.T.; Tsau, C.H.; Chang, S.Y. Nanostructured high-entropy alloys with multiple principal elements: Novel alloy design concepts and outcomes. *Adv. Eng. Mater.* **2004**, *6*, 299–311. [CrossRef]
3. Cantor, B.; Chang, I.; Knight, P.; Vincent, A. Microstructural development in equiatomic multicomponent alloys. *Mater. Sci. Eng. A.* **2004**, *375–377*, 213–218. [CrossRef]
4. Zhang, C.; Zhu, J.K.; Ji, C.Y.; Guo, Y.Z.; Fang, R.; Mei, S.W.; Liu, S. Laser powder bed fusion of high-entropy alloy particle-reinforced stainless steel with enhanced strength, ductility, and corrosion resistance. *Mater. Des.* **2021**, *209*, 109950. [CrossRef]
5. Prabhu, T.R.; Arivarasu, M.; Chodancar, Y.; Arivazhagan, N.; Sumanth, G.; Mishra, R.K. Tribological Behaviour of Graphite-Reinforced FeNiCrCuMo High-Entropy Alloy Self-Lubricating Composites for Aircraft Braking Energy Applications. *Tribol. Lett.* **2019**, *67*, 78–93. [CrossRef]
6. Muangtong, P.; Rodchanarowan, A.; Chaysuwan, D.; Chanlek, N.; Goodall, R. The corrosion behaviour of CoCrFeNi-x (x = Cu, Al, Sn) high entropy alloy systems in chloride solution. *Corros. Sci.* **2020**, *172*, 108740. [CrossRef]
7. Zhou, Y.J.; Zhang, Y.; Wang, Y.L.; Chen, G.L. Solid solution alloys of AlCoCrFeNiTi$_x$ with excellent room-temperature mechanical properties. *Appl. Phys. Lett.* **2007**, *90*, 181904. [CrossRef]
8. Moghaddam, A.O.; Sudarikov, M.; Shaburova, N.; Zherebtsov, D.; Zhivulin, V.; Solizoda, I.A.; Starikov, A.; Veselkov, S.; Samoilova, O.; Trofimov, E. High temperature oxidation resistance of W-containing high-entropy alloys. *J. Alloys Compd.* **2022**, *897*, 162733. [CrossRef]
9. Jin, B.Q.; Zhang, N.N.; Zhang, Y.; Li, D.Y. Microstructure, phase composition and wear resistance of low valence electron concentration Al$_x$CoCrFeNiSi high-entropy alloys prepared by vacuum arc melting. *J. Iron. Steel. Res. Int.* **2021**, *28*, 181–189. [CrossRef]
10. Campari, E.G.; Casagrande, A.; Colombini, E.; Gualtier, M.L.; Veronesi, P. The Effect of Zr Addition on Melting Temperature, Microstructure, Recrystallization and Mechanical Properties of a Cantor High-entropy Alloy. *Materials* **2021**, *14*, 5994. [CrossRef] [PubMed]
11. Nam, S.; Lee, H.W.; Jung, I.H.; Kim, Y.M. Microstructural Characterization of TiC-Reinforced Metal Matrix Composites Fabricated by Laser Cladding Using FeCrCoNiAlTiC High-entropy Alloy Powder. *Appl. Sci.* **2021**, *11*, 6580. [CrossRef]
12. Hruška, P.; Lukáč, F.; Cichoň, S.; Vondráček, M.; Čížek, J.; Fekete, L.; Lančok, J.; Veselý, J.; Minárik, P.; Cieslar, M.; et al. Oxidation of amorphous HfNbTaTiZr high-entropy alloy thin films prepared by DC magnetron sputtering. *J. Alloys Compd.* **2021**, *869*, 157978. [CrossRef]
13. Arif, Z.U.; Khalid, M.Y.; Rehman, E.U.; Ullah, S.; Tariq, A. A review on laser cladding of high-entropy alloys, their recent trends and potential applications. *J. Manuf. Process.* **2021**, *68*, 225–273. [CrossRef]
14. Fujieda, T.; Shiratori, H.; Kuwabara, K.; Kato, T.; Yamanaka, K.; Koizumi, Y.; Chiba, A. First demonstration of promising selective electron beam melting method for utilizing high-entropy alloys as engineering materials. *Mater. Lett.* **2015**, *159*, 12–15. [CrossRef]
15. Gao, P.; Fu, R.; Liu, J.; Chen, B.; Zhang, B.; Zhao, D.; Yang, Z.; Guo, Y.C.; Liang, M.X.; Li, J.P.; et al. Influence of Plasma Arc Current on the Friction and Wear Properties of CoCrFeNiMn High-entropy Alloy Coatings Prepared on CGI through Plasma Transfer Arc Cladding. *Coatings* **2022**, *12*, 633. [CrossRef]
16. Fan, Q.K.; Chen, C.; Fan, C.L.; Liu, Z.; Cai, X.Y.; Lin, S.B.; Yang, C.L. Ultrasonic suppression of element segregation in gas tungsten arc cladding AlCoCuFeNi high-entropy alloy coatings. *Surf. Coat. Technol.* **2021**, *420*, 127364. [CrossRef]

17. Duriagina, Z.; Kulyk, V.; Kovbasiuk, T.; Vasyliv, B.; Kostryzhev, A. Synthesis of Functional Surface Layers on Stainless Steels by Laser Alloying. *Metals.* **2021**, *11*, 434. [CrossRef]
18. Duriagina, Z.; Kovbasyuk, T.; Kulyk, V.; Trostianchyn, A.; Tepla, T. *Technologies of High-Temperature Insulating Coatings on Stainless Steels*; Engineering Steels and High Entropy-Alloys; IntechOpen: London, UK, 2020; pp. 57–80.
19. Shen, Q.K.; Kong, X.D.; Chen, X.Z. Fabrication of bulk Al-Co-Cr-Fe-Ni high-entropy alloy using combined cable wire arc additive manufacturing (CCW-AAM): Microstructure and mechanical properties. *J. Mater. Sci. Technol.* **2021**, *74*, 136–142. [CrossRef]
20. Dong, B.S.; Wang, Z.Y.; Pan, Z.X.; Muránsky, O.; Shen, C.; Reid, M.; Wu, B.T.; Chen, X.Z.; Li, H.J. On the development of pseudo-eutectic AlCoCrFeNi$_{2.1}$ high entropy alloy using Powder-bed Arc Additive Manufacturing (PAAM) process. *Mat. Sci. Eng. A–Struct.* **2021**, *802*, 140639. [CrossRef]
21. Fan, Q.K.; Chen, C.; Fan, C.L.; Liu, Z.; Cai, X.Y.; Lin, S.B.; Yang, C.L. Ultrasonic induces grain refinement in gas tungsten arc cladding AlCoCrFeNi high-entropy alloy coatings. *Mat. Sci. Eng. A–Struct.* **2021**, *821*, 141607. [CrossRef]
22. Sokkalingam, R.; Muthupandi, V.; Sivaprasad, K.; Prashanth, K.G. Dissimilar welding of Al$_{0.1}$CoCrFeNi high-entropy alloy and AISI304 stainless steel. *J. Mater. Res.* **2019**, *34*, 2683–2694. [CrossRef]
23. Shen, Q.K.; Xue, J.X.; Yu, X.Y.; Zheng, Z.H.; Ou, N. Triple-wire plasma arc cladding of Cr-Fe-Ni-Ti$_x$ high-entropy alloy coatings. *Surf. Coat. Technol.* **2022**, *443*, 128638. [CrossRef]
24. Cheng, J.B.; Sun, B.; Ge, Y.Y.; Hu, X.L.; Zhang, L.H.; Liang, X.B.; Zhang, X.C. Effect of B/Si ratio on structure and properties of high-entropy glassy Fe$_{25}$Co$_{25}$Ni$_{25}$(B$_x$Si$_{1-x}$)$_{25}$ coating prepared by laser cladding. *Surf. Coat. Technol.* **2020**, *402*, 126320. [CrossRef]
25. Chao, Q.; Guo, T.; Jarvis, T.; Wu, X.H.; Hodgson, P.; Fabijanic, D. Direct Laser Deposition Cladding of Al$_x$CoCrFeNi High-entropy Alloys on a High-temperature Stainless Steel. *Surf. Coat. Technol.* **2017**, *332*, 440–451. [CrossRef]
26. Middleburgh, S.C.; King, D.M.; Lumpkin, G.R.; Cortie, M.; Edwards, L. Segregation and migration of species in the CrCoFeNi high-entropy alloy. *J. Alloys Compd.* **2014**, *599*, 179–182. [CrossRef]
27. Zhang, G.J.; Tian, Q.W.; Yin, K.X.; Niu, S.Q.; Wu, M.H.; Wang, W.W.; Wang, Y.N.; Huang, J.C. Effect of Fe on microstructure and properties of AlCoCrFe$_x$Ni(x = 1.5,2.5) high-entropy alloy coatings prepared by laser cladding. *Intermetallics* **2020**, *119*, 106722. [CrossRef]
28. Wang, H.B.; Gee, M.; Qiu, Q.F.; Zhang, H.; Liu, X.M.; Nie, H.B.; Song, X.Y.; Nie, Z.R. Grain size effect on wear resistance of WC-Co cemented carbides under different tribological conditions. *J. Mater. Sci. Technol.* **2019**, *35*, 2435–2446. [CrossRef]
29. Jin, B.Q.; Zhang, N.N.; Yu, H.S.; Hao, D.X.; Ma, Y.L. Al$_x$CoCrFeNiSi high-entropy alloy coatings with high microhardness and improved wear resistance. *Surf. Coat. Technol.* **2020**, *402*, 126328. [CrossRef]
30. Archard, J.F. Contact and rubbing of flat surfaces. *J. Appl. Phys.* **1953**, *24*, 981–988. [CrossRef]
31. Liu, C.; Li, Z.; Lu, W.; Bao, Y.; Xia, W.Z.; Wu, X.X.; Zhao, H.; Gault, B.; Liu, C.L.; Herbig, M.; et al. Reactive wear protection through strong and deformable oxide nanocomposite surfaces. *Nat. Commun.* **2021**, *12*, 5518. [CrossRef]

Study on the Effect of Pulse Waveform Parameters on Droplet Transition, Dynamic Behavior of Weld Pool, and Weld Microstructure in P-GMAW

Jie Huang [1,2], Tao Chen [2], Daqing Huang [1,*] and Tengzhou Xu [2,*]

1. College of Electronic and Information Engineering, Nanjing University of Aeronautics and Astronautics, Nanjing 210016, China
2. College of Aeronautical Engineering, Nanjing Vocational University of Industry Technology, Nanjing 210023, China
* Correspondence: radiouav@nuaa.edu.cn or radiouav@sina.com (D.H.); 2018100938@niit.edu.cn (T.X.)

Abstract: The heating and impact of arc and droplet acting on the weld pool lead to the transfer of mass, heat, and momentum, which affects the dynamic behavior of the weld pool and the microstructure in the P-GMAW process. In this paper, an image processing program is used to extract the dynamic behavior characteristics of the droplet transition and the weld pool in high-speed photography. The influence of the current waveform on the arc pressure and the impact of the droplet is quantitatively analyzed with different parameters. The dynamic behavior of the weld pool and the microstructure under different current waveform conditions are further studied. The internal relation of current waveform parameters to weld pool behavior and weld microstructure was expounded. The results show that the droplet impact is positively correlated with the pulse peak current. The rectangular wave pulse has a more significant droplet impact than the exponential wave with the same waveform parameters. The impact of droplet transition on the weld pool enhances the convective intensity of the weld pool. It slows down the cooling rate of the solidified weld microstructure below the tail of the weld pool, increasing the grain size of the weld microstructure.

Keywords: P-GMAW; drop transition; weld pool; component supercooling

1. Introduction

The pulsed gas metal arc welding process (P-GMAW) is widely used in automated welding, which realizes the control of droplet transfer while ensuring welding efficiency. With the requirement for high welding quality in aerospace, transportation, and pressure vessel fields, the minute control of the welding process has become a new requirement of the manufacturing industry following the control of the droplet transfer process [1,2]. The microstructure of the weld is the dominant factor that affects the quality of the weld joint while ensuring the stability of the droplet transfer process [3,4]. The minute control of the welding process cannot be achieved only by optimizing the droplet transfer process. Many studies have shown that the weld pool's dynamic behavior affects the weld pool's solidification process and weld structure [5,6]. Therefore, the research on the dynamic behavior of the P-GMAW weld pool is of great significance to realizing minute control of the welding process. However, the mechanism of the influence of the P-GMAW current waveform on the dynamic behavior of the weld pool and the structure of the weld is still unclear, which hinders the further development of the P-GMAW.

The study of arc shape and droplet transition behavior under different waveform parameters of P-GMAW has important practical significance for optimizing pulse waveform. The drop transition forms in P-GMAW are mainly divided into three types: more drop in one pulse, one drop in one pulse, and one drop in more pulse, in which the transition form of one drop in more pulse is unstable and prone to short circuit resulting in splashes; while the weld of more drop in one pulse has finger-like penetration depth, which is easy to cause cracks. Many studies have proved that one drop in a vein is the best transition mode of droplet melting [7,8]. The current pulse waveform also affects the welding wire and weld pool's heat and mass transfer process. Existing studies have shown a strong correlation between the momentum carried by droplets when they impact the weld pool and the weld penetration depth [9,10]. In literature, the concept of compelling momentum in the droplet transition process was introduced and defined as the ratio of the total momentum carried by the droplet in unit time to the welding speed, which could better characterize the influence of the droplet speed on the weld depth. The adjustment of the current waveform significantly affects the heat transfer process acting on the weld pool and the cooling time of the welded joint. While controlling the welding heat input, the current waveform also changes the heat conduction process [2], significantly affecting the weld pool's solidification process and the weld microstructure [7]. Many studies have shown that the P-GMAW welding process can adjust the weld metal structure by adjusting the pulse frequency and droplet impact strength [11,12]. The impact of the current pulse acting on the weld pool will lead to the breakage of trace solidified dendrites and the presence of equiaxial grains in the molten pool, which are conducive to the nucleation of dendrites and grow into equiaxial primary phase, and finally, define the weld structure.

A large number of research results have been obtained on the droplet transition process and arc of the P-GMAW process, but research on the current waveform is only limited to the droplet transition process. More research on the influence mechanism of the current waveform on the dynamic behavior of weld pool and weld microstructure could be learned. In summary, existing research has the following deficiencies:

(1) Existing studies have analyzed the influence of current waveforms on droplet transition behavior. However, there is no quantitative analysis of the heat input and impact of arc and droplet transition acting on the weld pool surface, so the differences in heat and force effects on the weld pool under different current waveforms cannot be quantitatively compared.
(2) The influence of P-GMAW current waveform on weld pool oscillation behavior, the solidification process of weld pool, and weld microstructure are rarely studied and reported systematically.

In this paper, for the P-GMAW, the effect of the current waveform on the heat and force acting on the weld pool, the weld pool behavior, the weld microstructure, and properties are based on the welding high-speed photography system and image processing system.

This paper quantitatively analyzed the differences in heat, force action, and dynamic behavior of weld pools under different pulse peak currents and pulse waveform shapes based on the self-developed high-speed welding photography platform and image processing system. At the same time, combined with the flow behavior of the weld pool, the influence of the current waveform on the solidification process of the weld pool and the weld microstructure was revealed in order to study the influence of the current waveform on the dynamic behavior of the weld pool and the weld microstructure of the P-GMAW process systematically.

2. Materials and Methods

The welding high-speed photography system used in this paper is shown in Figure 1. In order to capture images of the weld pool from different angles, the high-speed camera shot the welding area from two angles: highspeed camera 1 was placed horizontally, and the arc morphology, droplet transition process, and the profile of the upper surface of the molten pool could be observed. Highspeed camera 2 was used to observe the flow behavior of the weld pool surface. The high-speed camera is fixed relative to the welding torch, and the shooting direction is perpendicular to the weld. The electrical signal acquisition system in the welding process is composed of a current Hall sensor (model AKHC-EKAA DC), a voltage Hall sensor (model KCE-VZ01), and a signal acquisition card (model NI 6251, collection frequency 5×10^5 Hz).

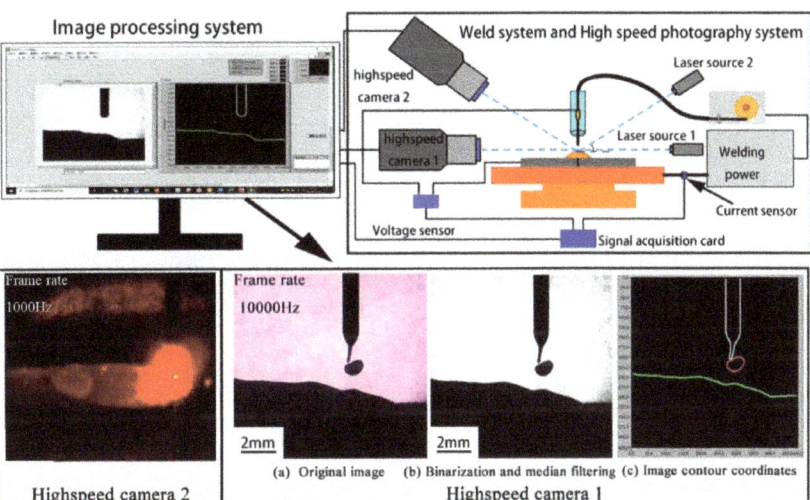

Figure 1. Experimental system and image processing system.

The wavelength of the laser source 1 is 850 nm, and the 850 nm narrow-band filter is attached to the high-speed camera 1. After the software processing of the high-speed photography pictures taken by highspeed camera 1, the contour coordinates of the molten droplets and weld pools in the pictures can be obtained, as shown in Figure 1.

In P-GMAW, alloying elements such as silicon and manganese in the base metal and the wire had a high affinity to react with oxygen and form silicon oxide and manganese oxide. These oxides accumulate on the surface of the weld pool and form slag [13]. The slags have a lower density than the molten metal and follow the flow pattern of the weld pool; hence, the slag flow pattern and accumulation location can disclose the weld pool flow behavior [14]. The torch angle of highspeed camera 2 is 60°. In order to capture the outline of slags (the oxide on the surface of weld pool) and avoid the interference of arc light, the laser source and narrowband filter with a wavelength of 650 nm was selected to highlight the outline information of slags on the weld pool, as shown in Figure 1.

In order to study the influence of pulse current waveform parameters on droplet size and droplet velocity, the center of gravity, diameter, and droplet velocity of the droplet should be measured by image processing technology. A rectangular coordinate system can be set up according to the arrangement of pixels on the high-speed photography image. The x-axis is the bottom edge of the image, the direction is left, the y-axis is the left side of the image, and the direction is up. Since the welding gun and wire are perpendicular to the workpiece surface, and the camera lens is perpendicular to the weld, the speed of the dropper perpendicular to the picture surface (in the z-dimension direction) can be ignored. Due to the effect of surface tension on the droplet, the droplet will approximate into a sphere. For convenient analysis and calculation, the droplet is regarded as a sphere. In addition, the size of the picture taken by the high-speed camera is 480 × 428, and the diameter of the welding wire is 1.2 mm. The pixel size can be quantified by measuring the diameter of the welding wire in the picture while maintaining the position of the camera and torch. The extraction process is as follows:

(1) The contours and positions of the droplets at different times during the welding process were extracted, as shown in Figure 1. The droplet area ($S_{droplet}$), the coordinates of droplet center of gravity (X_{Gt}, Y_{Gt}), and droplet diameter ($D_{droplet}$) were calculated based on the contour coordinates of the droplet, as shown in Equations (1)–(3).

$$S_{droplet} = \sum_{x=x_{min}}^{x_{max}} (y_{max} - y_{min}) \qquad (1)$$

$$(x_{Gt}, y_{Gt}) = \left(\frac{\sum_1^n x_{nt}}{n}, \frac{\sum_1^n y_{nt}}{n} \right) \tag{2}$$

$$D_{droplet} = \sqrt{\frac{4 \times S_{droplet}}{\pi}} \tag{3}$$

where (X_{nt}, Y_{nt}) is all coordinates of the droplet contour in Figure 1, (X_{max}, Y_{max}) is the maximum coordinate value of the droplet contour, (X_{min}, Y_{min}) is the minimum coordinate value of the droplet contour, and n is the number of droplet contour points. $S_{droplet}$ is the area of the droplet in high-speed photography and $D_{droplet}$ is the equivalent diameter of the droplet.

(2) The coordinates of the center of gravity position of the droplet are continuously collected, as shown in Figure 2. The droplet velocity is calculated according to Equation (4), and the droplet momentum is calculated according to Equation (5).

$$V_{droplet} = \frac{\sqrt{(x_{Gt1} - x_{Gt2})^2 + (y_{Gt1} - y_{Gt2})^2}}{|t_2 - t_1|} \tag{4}$$

$$M_{droplet} = \frac{\pi D_{droplet}^3}{6} \rho V_{droplet} \tag{5}$$

where $V_{droplet}$ is the velocity of the droplet, ρ is the density of the liquid metal, and $P_{droplet}$ is the momentum of the droplet.

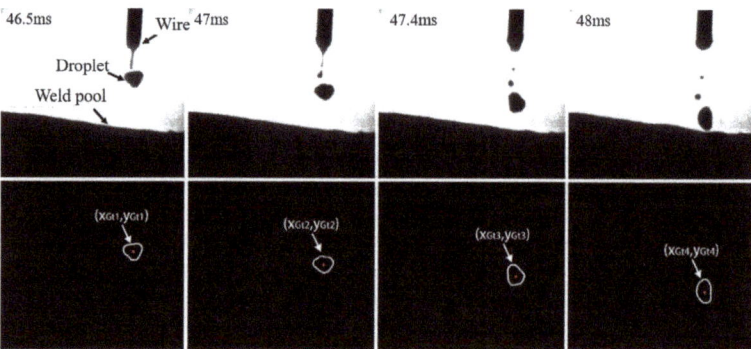

Figure 2. The droplet center of gravity position at a different time during P-GMAW welding.

In this paper, Fronius TPS5000 and Megmeet PM500A were selected as power sources of the welding system to provide the needed waveform. Fronius TPS5000 is used to provide a trapezoidal pulse waveform, and Megmeet PM500A is used to provide an exponential pulse waveform. The pulse waveform parameters are automatically set by the unified control system of the welding power source according to the welding current, as shown in Figure 3. I_p is peak pulse current, I_b is Pulse base value current, and I_m is mean current. In order to ensure the stability of one pulse and one drop, the pulse width provided by both welding power sources is 4 ms.

The I_p of the trapezoidal pulse waveform supplied from a Fronius TPS5000 can be independently adjusted. We used 4 mm Q235 low carbon steel and 1.2 mm ER50-6 carbon steel wire, respectively, as the base metal and welding wire. The shielding gas is 82% Ar + 18% CO_2 mixture, and the gas flow rate is 20 L/min. The welding speed with I = 60 A is 25 cm/min, the welding speed with I = 100 A is 38.5 cm/min. The process parameters are shown in Table 1. The other pulse waveform parameters were automatically set by the unified control system of the welding power supply according to the welding current.

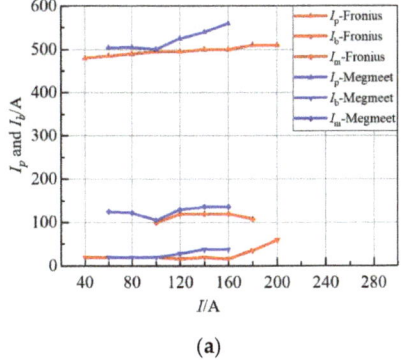 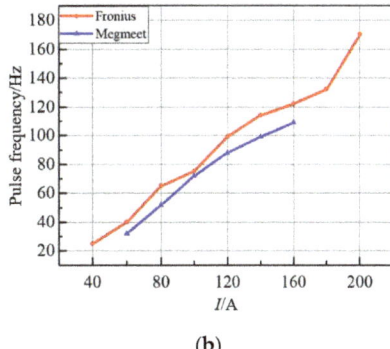

(a) (b)

Figure 3. The pulse waveform parameters with different currents supplied by a Fronius TPS5000 and Megmeet PM500A: (**a**) parameters of the waveform; (**b**) frequency of the pulse.

Table 1. The welding parameters used in this article.

	Waveform	I/A	I_p/A	I_b/A
1	Trapezoidal type	60	485–525	20
2	Exponential type	100	500	20
3	Trapezoidal type	40–200	Set automatically	
4	Exponential type	60–180	Set automatically	

The arc heat and the heat carried by the droplet are the main heat source of the weld pool in P-GMAW. It is assumed that the current density is evenly distributed on the projection surface of the arc and weld pool. The arc heat ($E_{arc-pulse}$) acting on the weld pool during a single current pulse is shown in Equation (6), and the heat carried by the molten droplet ($E_{droplet-pulse}$) is shown in Equation (7). The total energy input ($E_{total-pulse}$) on the surface of the molten pool for a single pulse cycle is shown in Equation (8), and the energy input power ($P_{total-pulse}$) on the surface of the molten pool is shown in Equation (9).

$$E_{arc-pulse} = \int_0^{1/f} P_{arc} dt = \int_0^{1/f} I(V_w - \varphi) dt \tag{6}$$

$$E_{droplet-pulse} = \rho \frac{4}{3}\pi \left(\frac{D_d}{2}\right)^3 \int_{300}^{2500} C_p dT \tag{7}$$

$$E_{total-pulse} = \rho \frac{4}{3}\pi \left(\frac{D_d}{2}\right)^3 \int_{300}^{2500} C_p dT + \int_0^{1/f} I(V_w - \varphi) dt \tag{8}$$

$$P_{total-pulse} = E_{total-pulse} \times f \tag{9}$$

where f is the pulse current frequency; V_w is the cathode pressure drop, which is 16.7 V [8]; φ is the electron escape work of the base metal, which is 4.77 V [10]; ρ is the density of the liquid metal, D_d is the equivalent diameter of the molten droplet, C_p is the specific heat capacity of the liquid metal, and T is the molten drop temperature, assumed to be 2500 °C.

The arc force acting on the weld pool (p_{arc}) during P-GMAW welding is shown in Equation (10), and the equivalent pressure ($p_{droplet}$) of the droplet impact on the weld pool is shown in Equation (11).

$$p_{arc} = \frac{\mu}{4\pi} I^2 \log\left(\frac{D_P}{D_R}\right) / \frac{\pi D_p^2}{4} \tag{10}$$

$$p_{droplet} = M_{droplet} f / \frac{\pi D_d^2}{4} = \frac{2 D_d \rho V_d f}{3} \tag{11}$$

where μ is the spatial permeability, $\mu = 1.26 \times 10^{-6}$ N/A^2; ρ is the density of the liquid metal, $\rho = 7.85$ g/cm^3; f is the pulse frequency, $f = 40$ Hz; V_d is the droplet velocity; D_d is the droplet equivalent diameter; D_P is the projected diameter of arc on the weld pool. D_R is the diameter of the arc root. The dimension parameters of the arc in the welding process are shown in Figure 4.

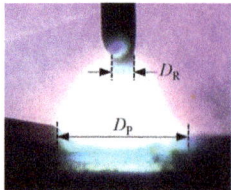

Figure 4. Schematic diagram of arc profile dimensions.

3. Results and Discussion

3.1. Effect of Current Waveform on Heat and Force Acting on Welding Pool

3.1.1. Effect of I_p on Heat and Force Acting on Welding Pool

The current waveform of the P-GMAW process is periodic, and the thermal and force effects on the welding pool during the whole welding process can be analyzed by studying the droplet and arc behavior in a single pulse period. Figure 5a,b are the current and voltage waveforms with I_p of 485 A and 525 A, respectively, and the welding current is both 60 A.

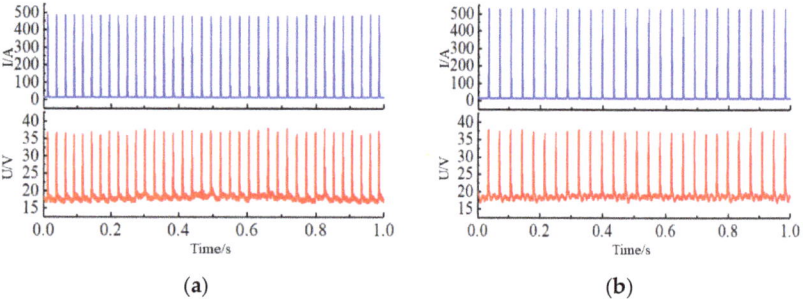

Figure 5. The current and voltage waveform in P-GMAW: (a) $I_p = 485$ A, $I = 60$ A; (b) $I_p = 525$ A, $I = 60$ A.

Figures 6 and 7 show the high-speed photography of the arc and droplet transition process and synchronous welding electrical signals within a single current pulse period with the parameters of $I_p = 485$ A and $I_p = 525$ A, respectively. It can be found that the arc size increases and the arc brightness increases at the peak pulse stage, and the welding wire melts rapidly under the action of arc heat and forms a molten droplet hanging at the end of the wire. In the late pulse peak stage, the molten droplet falls off, and the current at the pulse base value stage falls back to the base value. The function of the current at the base value is to maintain the arc, which is 20 A, and the arc is almost invisible. In the whole pulse period, there was no obvious melting at the end of the wire at the base value stage.

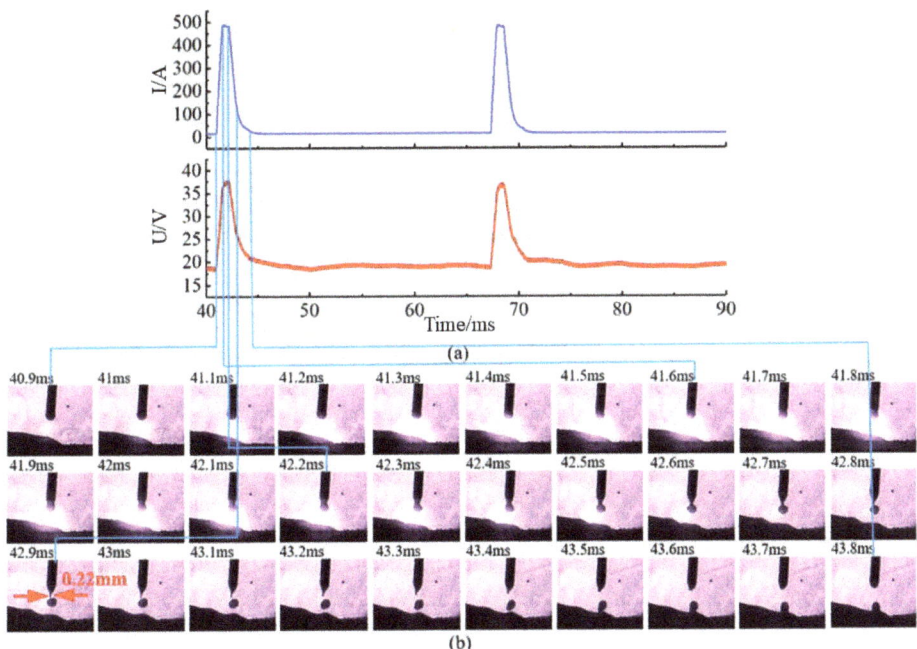

Figure 6. P-GMAW current and voltage signal and arc and droplet transition behavior photographs (I_p = 485 A): (**a**) current and voltage signal; (**b**) synchronous photographs.

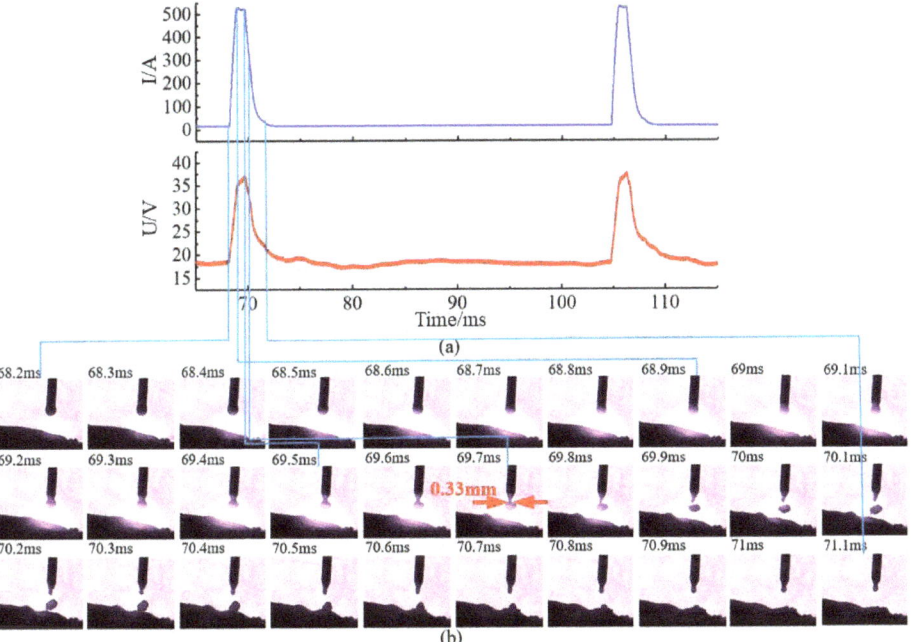

Figure 7. P-GMAW current and voltage signal and arc and droplet transition behavior photographs (I_p = 525 A): (**a**) current and voltage signal; (**b**) synchronous photographs.

The increase of I_p lead to the decrease of pulse frequency with the same average current (under the condition of 60 A welding current, I_p = 485 A, f = 40 Hz, I_p = 525 A, f = 30 Hz), and higher I_p accelerate the melting droplet from the welding wire. As shown in Figures 6 and 7, it takes 2 ms from the beginning of the pulse to the moment when the droplet is released from the wire with I_p = 485 A, and the shrinking neck diameter before the droplet is 0.22 mm, while it takes 1.6 ms with I_p = 525 A, and the shrinking neck diameter before the droplet is 0.33 mm. The results show that with the higher I_p, the shorter the time it takes, and the larger the diameter of the shrinking neck as the droplet to detangle from the wire.

Figure 8a shows the droplet and arc size at different times during a single pulse period (I_p = 485 A). The change process of heat and force acting on the weld pool can be calculated by combining Equations (6) and (11) with the current and voltage data in Figure 4, as shown in Figure 8b. The D_P is synchronized with the current, however, at the moment when the droplet detaches from the wire, a peak occurs in the D_P curve, which is due to the increase of metal vapor concentration in the arc space resulting from by the droplet detaching from the wire. The D_R Curve first falls and then rises at the peak pulse stage because the arc root is always located at the shrink neck between the welding wire and the droplet. The droplet gradually grows and forms a shrink neck, and the size of the arc root decreases gradually. As the droplet detaches, the shrink neck disappears, and the size of the arc root increases. The arc thermal power also has the same variation trend as the current curve, as shown in Figure 8b. The arc pressure increases rapidly with the rise of the current and presents a brief stable stage similar to the current waveform after reaching the peak value. The peak arc pressure is nearly 580 Pa. With the current falling to the base value, the arc pressure falls back and maintains at nearly 5 Pa. The impact of the droplet acting on the weld pool at a fixed frequency can be equivalent to equivalent force, as shown in Equation (11). However, the solidification process of the weld pool is a phase transition process of rapid cooling, and the equivalent force cannot accurately measure the effect of a single molten droplet impacting on the weld pool. Therefore, the momentum carried by a single molten drop is another major parameter to measure the impact. When I_p = 485 A, the projection area of the droplet is 1.1 mm^2, the velocity of the droplet is 1.09 m/s, the momentum carried by the droplet is 0.69×10^{-5} kg·m/s, and the equivalent pressure of the droplet impact on the weld pool is 251 Pa. The welding heat input power is 2.41×10^3 J/s.

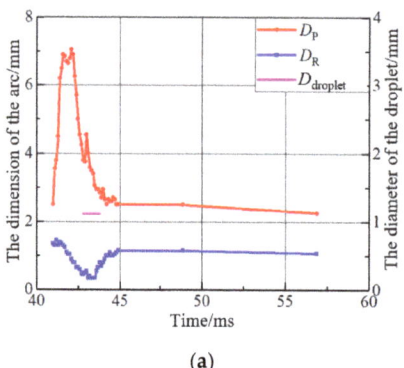
(a)

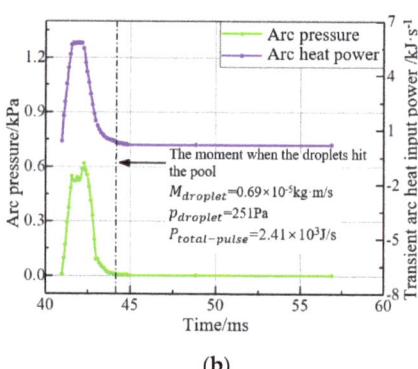
(b)

Figure 8. (a) The droplet and arc size at different times during a single pulse period (I_p = 485 A); (b) transient arc pressure and heat input power (I_p = 485 A).

Figure 9 shows the size of the molten droplet and arc and the thermal and force effects on the weld pool with the I_p = 525 A. The arc size increased significantly as the peak pulse current increased, and the arc pressure was maintained at about 610 Pa during the peak pulse phase, which was significantly higher than the peak pulse pressure with the I_p = 485 A, the projected area of the droplet is 1.15 mm^2, and the velocity is 1.65 m/s. Due to the increase of heat input in the peak phase of the pulse, the size and velocity of the droplet increase simultaneously, which leads to the rise of the droplet momentum. According to the calculation, As I_p increased to 525 A, the momentum carried by the droplet increased to 1.20×10^{-5} kg·m/s, and the equivalent impact force of the droplet acting on the weld pool is 313 Pa. At the same time, a too-high pulse peak current could increase the volume of residual droplets at the end of the welding wire, which can easily lead to the occurrence of multi drop during one pulse, as shown in Figure 10. Higher I_p increases the energy input of a single current pulse, while reducing the pulse frequency, resulting in no significant change in the total heat input power during the welding process. According to the measurement, the total heat input power received by the weld pool as the I_p is 525 A is 2.38×10^3 J/s, which is approximately it as the I_p is 485 A.

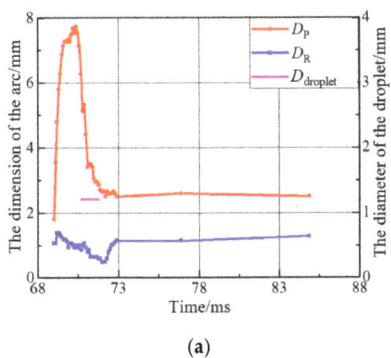
(a)
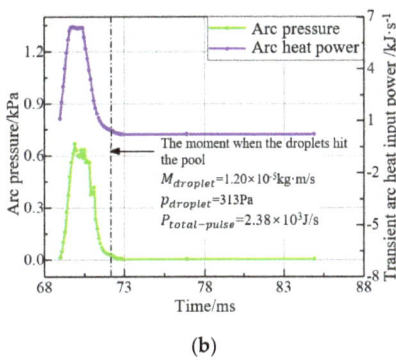
(b)

Figure 9. (a) The droplet and arc size in different times during a single pulse period (I_p = 485 A); (b) transient arc pressure and heat input power (I_p = 525 A).

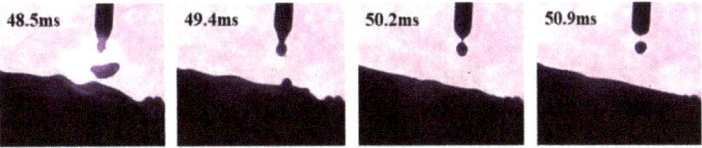

Figure 10. The phenomenon of multi-drop during one pulse (I_p = 525 A).

Figure 11 shows the influence of pulse peak current on arc pressure, droplet momentum, and equivalent impact force when the welding current is equal to 60 A. It can be found that with the increase of I_p, the equivalent impact force and arc pressure of the molten drop rise slowly, but the momentum of the molten droplet increases significantly (when I_p increases from 485 A to 525 A, the momentum of the molten droplet rises nearly 200%).

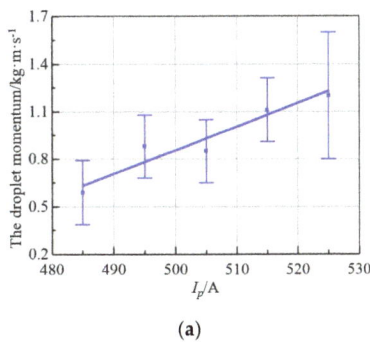
(a)
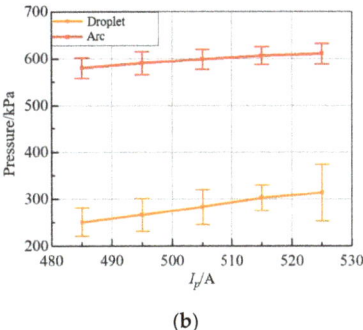
(b)

Figure 11. Effect of I_p on droplet momentum, arc, and droplet impact pressure. (a) The droplet momentum; (b) the pressure of the arc and droplet.

At the peak of the pulse, the electromagnetic contraction force generated by the current passing through the droplet is one of the leading forces driving the droplet off the wire, which affects the initial velocity of the droplet when it leaves the wire [15]. At the same time, the metal flow of the molten droplet at the end of the welding wire will produce a downward force F_{mf}, which could increase the initial velocity of the droplet [16]. The droplet accelerates under the impact of the high-speed plasma current in the arc space after the droplet disconnects from the wire and enters the arc space [17]. The downward force F_{mf} and the axial force F_a on the molten droplet in the arc space increase with the current increase [15–17]. It can be proved that the force received by the molten drop is higher before it disconnects from the wire as the I_p increase, which explains that the higher the peak pulse current, the shorter the time taken for the molten droplet to disconnect from the wire, and the larger the diameter of the shrinking neck before disconnecting. At the same time, according to references [15–17], higher I_p could lead to a more significant initial velocity of the molten droplet when it breaks away from the welding wire. At the same time, it will be affected by more intense plasma impact in the arc space, and the acceleration of the molten droplet will increase, leading to the increase of the momentum of the molten droplet. Due to the decrease of in pulse frequency, the increase of droplet impact force is slower than that of droplet momentum, which leads to the equivalent force of droplet impact is significantly smaller than that of arc pressure. However, the difference in the impact form of the arc and droplet on the weld pool does not mean that the impact of the droplet on the weld pool is much smaller than that of the arc. At the same time, it has been shown that the arc force at the peak of the pulse cannot significantly change the surface state of the weld pool. The impact of the molten droplets is the main factor causing the oscillation behavior of the weld pool [18]. However, I_p is positively correlated with the impact of the weld pool.

3.1.2. Effect of Waveform Shape on Heat and Force Acting on Welding Pool

Due to the high coupling between welding parameters and pulse waveform parameters, there is no apparent linear relationship between waveform parameters of one pulse and one drop under different current conditions. Therefore, many welding power supply manufacturers have established welding parameter libraries to ensure the stability of one pulse and one drop, which leads to differences in pulse waveforms under the same current. When the welding current is between 100 A and 180 A, the preset pulse waveform of the FroniusTPS5000 changes from trapezoid to posterior median trapezoid. The preset pulse waveform of the Megmeet PM500A is the posterior median exponential pulse at the full range of welding currents. Many scholars have thoroughly studied the effect of the posterior median parameters of the pulse current waveform on the melting droplet transition process [19]. The results show that the central role of the posterior median pulse is to supplement the pulse's peak phase energy and increase the pulse-drop transition's stability [19]. Therefore, this section will not analyze the effect of the median current after the pulse on the melt transition process.

In order to avoid the influence of the difference in pulse waveform parameters with a different welding power source, this section carries out the test at the intersection point of the process parameters of the two pulse waveforms. When the welding current is 100 A, the trapezoid wave parameters (I_p = 500 A; I_b = 17 A; I_m = 100 A; f = 75 Hz) are similar to exponential wave parameters (I_p = 500 A; I_b = 20 A; I_m = 100 A; f = 73 Hz), as shown in Figure 3. The arc and droplet transition behavior of trapezoidal pulse and exponential pulse are shown in Figures 12 and 13.

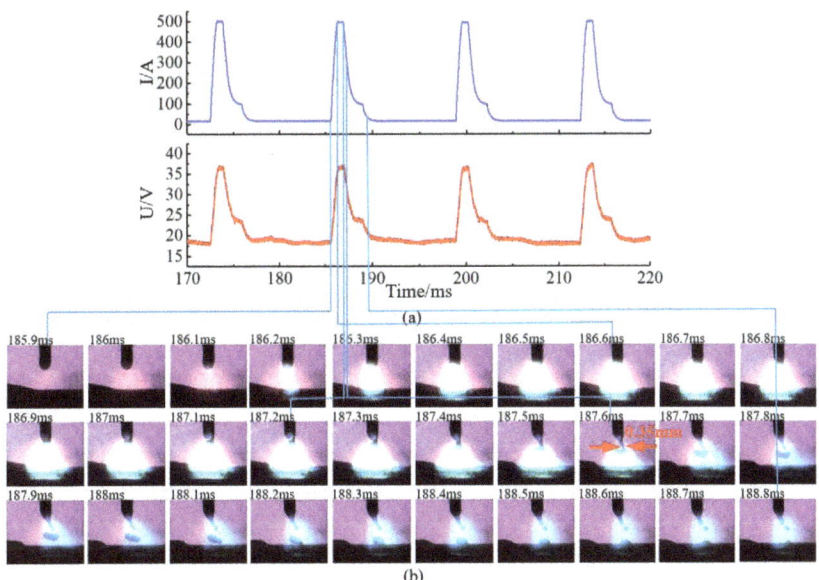

Figure 12. P-GMAW electronic signal and arc and droplet transition behavior photographs (trapezoid pulse waveforms, I = 100 A): (**a**) current and voltage signal; (**b**) synchronous photographs.

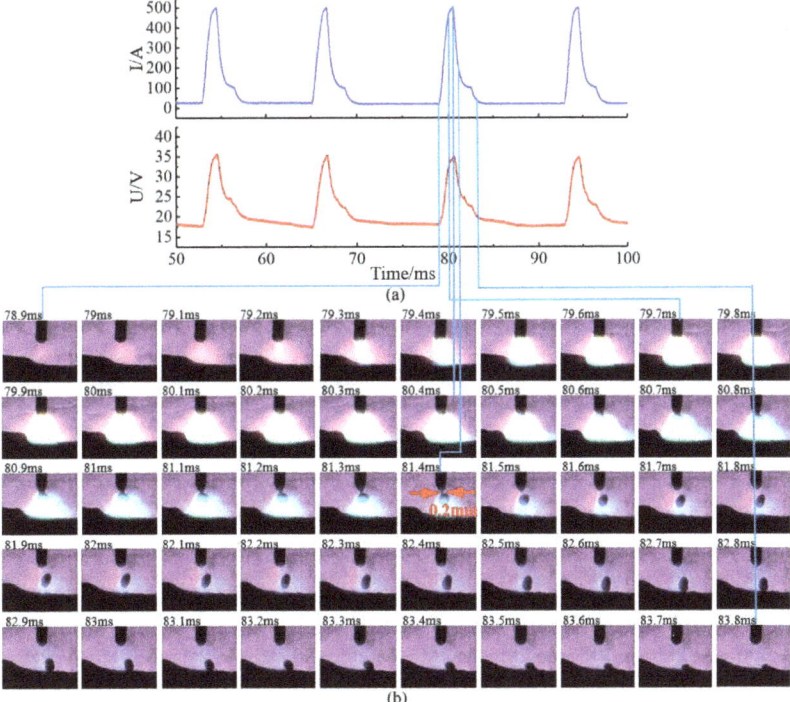

Figure 13. P-GMAW current and voltage signal and arc and droplet transition behavior photographs (exponential pulse waveform, I = 100 A): (**a**) current and voltage signal; (**b**) synchronous photographs.

The pulse onset time of the trapezoid pulse waveform was 185.9 ms, and the droplet detachment time was 187.7 ms. The time from the pulse onset time to the moment of droplet detachment from the wire was 1.8 ms, and the diameter of droplet retraction before detachment from the wire was 0.35 mm. The onset time of the exponential pulse waveform was 78.9 ms, the time of droplet detachment was 81.5 ms, the time of droplet detachment was 2.6 ms, and the diameter of the shrinking neck before droplet detachment was 0.22 mm. It took longer for the droplet of the exponential pulse to leave the wire than that of the trapezoidal pulse, and the diameter of the shrinking neck before the droplet leaving the wire was significantly smaller.

Under the same pulse peak current, the droplet in the arc space of the trapezoidal pulse was significantly compressed, while the droplet of the exponential pulse had no obvious deformation. The reason for the above differences is that the current rises quickly at the initial stage of the trapezoid pulse and the metal vapor concentration in the arc space is relatively small, forcing the arc to climb to the upper end of the melt drop, and the electromagnetic contraction force rises faster, as shown in Figure 12b 187.2–187.8 ms, which further enhances the extrusion effect of the electromagnetic contraction force on the constricted neck, and the melt drop is easy to break off the welding wire. The slow rise rate of the exponential pulse waveform current leads to a relatively sufficient metal vapor in the arc space, which allows a larger current to pass through the bottom of the melt drop, resulting in the arc concentrated below the melt drop and did not climb to the position of the shrinking neck, which to a certain extent prevents the separation of the melt drop, as shown in Figure 13b 80.8–81.5 ms. The above phenomena are similar to the results obtained by Hertel through simulation [20].

The droplet projection size of the trapezoidal pulse waveform was 1.156 mm^2, and the velocity was 1.28 m/s. The droplet projection size of the exponential pulse waveform was 1.09 mm^2, and the velocity was 0.82 m/s. Previous studies have proven that the increase of I_p could lead to the simultaneous increase of electromagnetic force and high-speed convection intensity in the droplet and then increase the initial velocity of the droplet [20]. Under the same I_p condition, the trapezoid pulse takes less time for the droplet to detach from the wire. The current is 300 A when the droplet detangles from the wire of the trapezoid pulse, which is significantly higher than that of the exponential pulse (the current is around 200 A when the droplet detaches from the wire), which results in a stronger driving effect on the arc of the trapezoid pulse after the droplet detangles from the wire (as shown in Equation (14)). As a result, the droplet has a large acceleration and initial velocity [21–24].

The droplet and arc sizes of the two waveforms in a single pulse cycle are shown in Figures 14a and 15a. Combined with Equations (6)–(11), the heat and force acting on the weld pool in a single pulse cycle can be calculated, as shown in Figures 14b and 15b. It can be found that the welding heat input power of the trapezoidal pulse and exponential pulse are similar, and the maximum value of arc size is close. However, the increase rate of arc size of the trapezoidal wave is significantly faster than that of the exponential wave, and the change of arc pressure shows a similar trend with the arc size. Although the peak arc pressure of the two waveforms is similar, the increased rate of arc size of the trapezoidal wave is significantly higher than that of the exponential wave due to the higher current rise rate of the trapezoidal pulse. As a result, its arc pressure rises faster, and its residence time above 600 Pa is longer than that of the exponential pulse. The droplet velocity of the trapezoidal pulse is significantly higher than that of the exponential pulse, and the droplet momentum and impact force of the trapezoidal pulse is higher than that of the exponential pulse.

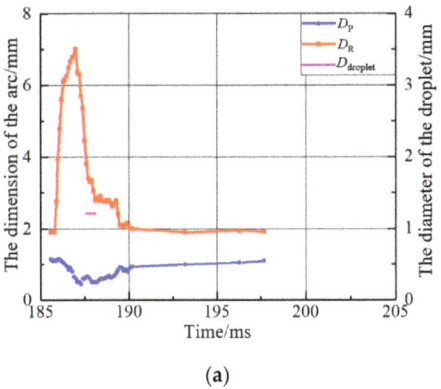 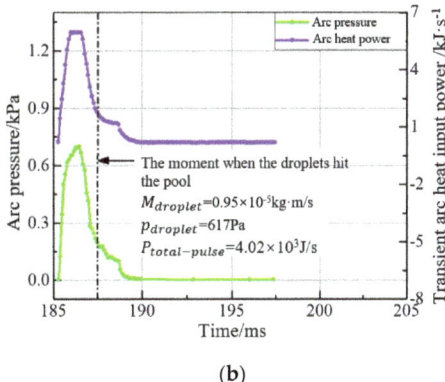

(a) (b)

Figure 14. (a) The droplet and arc size in different times during a single pulse period (exponential pulse waveform, I = 100 A); (b) transient arc pressure and heat input power (exponential pulse waveform, I = 100 A).

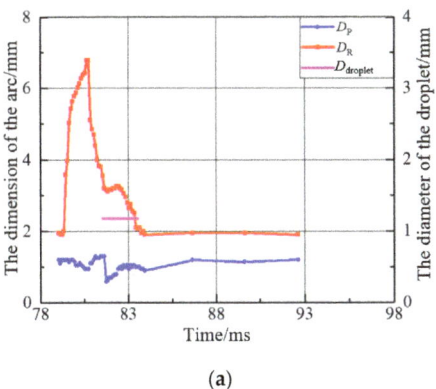 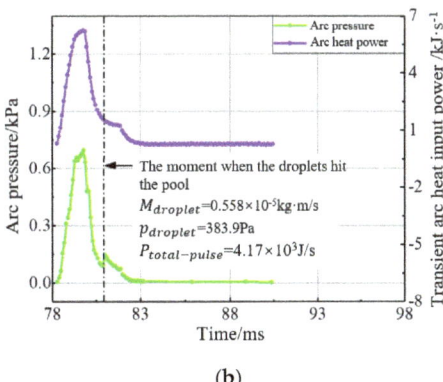

(a) (b)

Figure 15. (a) The droplet and arc size at different times during a single pulse period (trapezoid pulse waveform, I = 100 A); (b) transient arc pressure and heat input power (trapezoid pulse waveform, I = 100 A).

Figure 16 shows the variation trend of droplet momentum, equal impact force, and arc pressure of trapezoidal pulse and exponential pulse under different welding current conditions. According to Figure 16a–c, it can be found that the momentum and the droplet impact equivalent pressure of trapezoidal pulse are higher than those of exponential pulse under all current welding parameters. As shown in Figures 3a and 16a, with the increase of welding current, the peak pulse current also increases, which leads to the synchronous increase of droplet speed and droplet quality, and the above factors all increase the momentum of the droplet. Figure 16d shows the peak pressure of the arc under different welding current conditions. Combined with the distribution of I_p in Figure 3a, it can be found that the peak pressure of the arc is positively correlated with I_p and has no significant correlation with the pulse waveform shape.

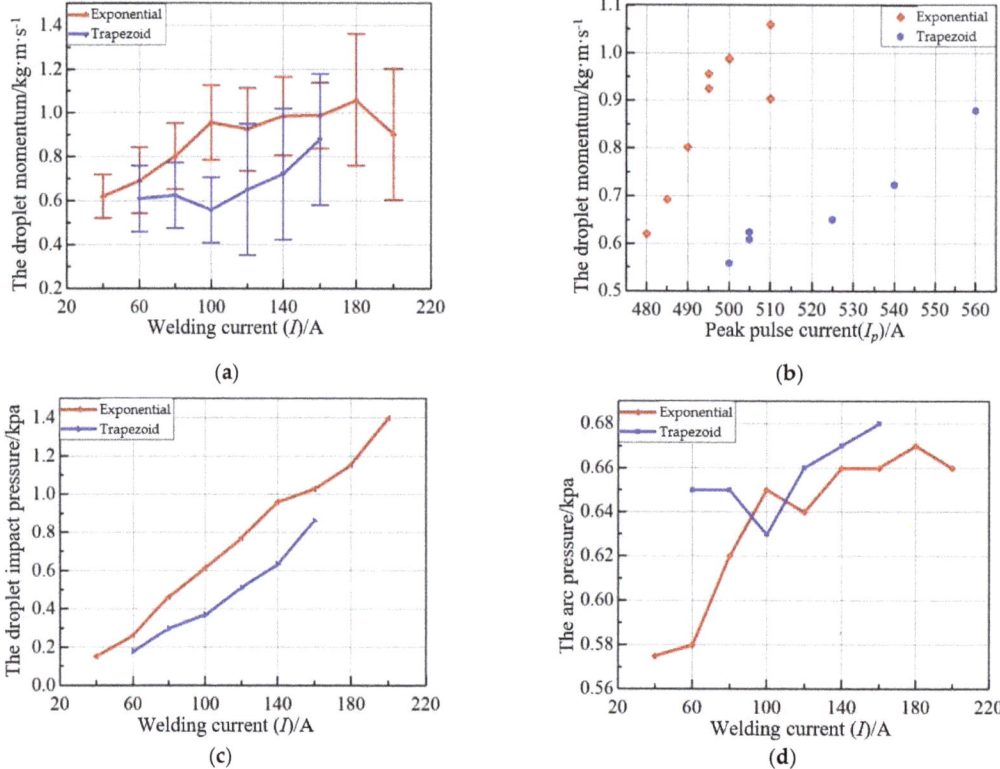

Figure 16. (a) Droplet momentum with different welding current; (b) droplet momentum with different I_p; (c) the droplet impact equivalent pressure with different welding current; (d) the arc pressure with different welding current.

3.2. Effect of Current Waveform on the Welding Pool Dynamic Behavior and Microstructure

3.2.1. Effect of Current Waveform on the Dynamic Behavior of Welding Pool

In the welding process, Si, Mn, and other alloying elements in the base metal and welding wire have a high affinity with O elements in the protective gas, and SiO_2 and MnO substances formed by oxidation of alloying elements float on the surface of the welding pool to form slag. The density of slag is lower than that of liquid metal, and researchers have shown that the movement behavior and accumulation position of molten slag reveals the flow behavior of the weld pool [25].

Figure 17 shows the high-speed photography of welding pool flow behavior and molten slag morphology with the I_p 485 A and 525 A, respectively. Figure 18 shows exponential pulse waveforms and trapezoid pulse waveforms with I = 100 A. According to the movement behavior of molten slag, a schematic diagram of the convection behavior of the molten pool can be deduced, as shown in Figure 19. After the front end of the welding pool is compressed, the metal flow is blocked by the bottom and forced to flow to the rear of the welding pool. After being rebounded by the rear edge, it moves to the front end of the welding pool. At the same time, the metal flow on the front surface of the welding pool will be pushed to the rear under the impact and surface tension, and the two metal flows will converge at the rear of the welding pool, resulting in the aggregation of metal oxides into slag blocks.

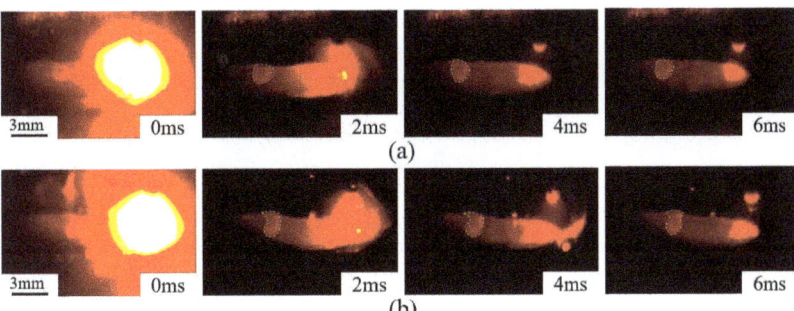

Figure 17. Effect of I_p on the flow process of the welding pool and the morphology of molten slag in P-GMAW process: (**a**) I_p = 485 A; (**b**) I_p = 525 A.

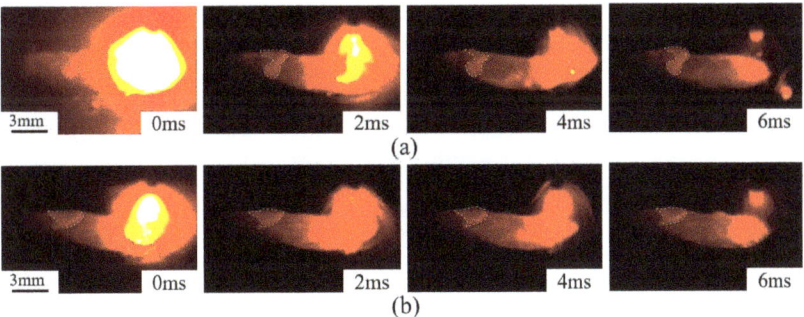

Figure 18. Effect of pulse waveform shape on the flow process of the welding pool and the morphology of molten slag in P-GMAW process: (**a**) exponential pulse waveform; (**b**) trapezoid pulse waveform.

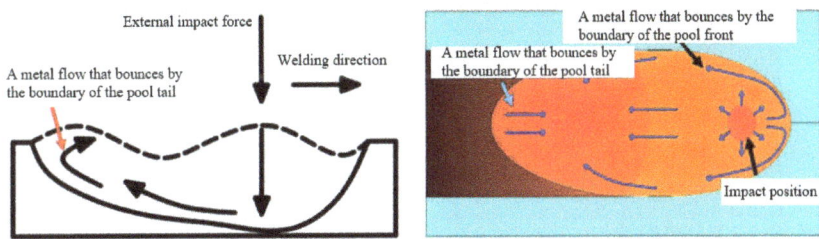

Figure 19. Schematic diagram of welding pool flow behavior.

As shown in Figure 17, the welding pool with I_p = 485 A has a whole block of circular molten slag on its surface, and the welding pool surface with I_p = 525 A has a noticeable depression in the center of the molten slag.

As shown in Figure 18, The middle part of the molten slag on the surface of the trapezoidal pulse pool shrinking obviously, showing a typical "barbell" shape, while the molten slag of the exponential pulse gathered in a circle at the end of the welding pool.

The depression of the slag center indicates that it is subjected to strong convection extrusion in the welding pool. According to the content of Section 3.1, the droplet velocity is positively correlated with the convection intensity of the welding pool. Under the same parameters, the larger the droplet velocity, the higher the kinetic energy transferred during the momentum exchange of the impact welding pool, which leads to an increase in the convection intensity of the welding pool.

3.2.2. Effect of Current Waveform on the Solidification Process and Weld Microstructure

In the welding process, the front end of the welding pool is affected by the heat and force of the arc and the droplets, while the liquid metal at the rear of the welding pool solidifies continuously. Liquid metal is always in a complex state of motion. In order to better understand the influence of the current waveform on the solidification process of the molten pool, it is necessary to calculate the solidification rate of the welding pool (R), the cooling rate of the weld (C_R) and the temperature gradient (G_S and G_L).

In the welding process, the liquid metal at the back edge of the welding pool solidifies rapidly, and the growth direction of the grains is related to the shape of the welding pool and the welding speed. In general, the grain of the weld is oriented in the welding direction and bent to the weld center, and the grain's growth rate (R) is the linear speed of the trailing edge of the welding pool. Under the steady state and quasi-steady state welding conditions, the growth rate of columnar crystals at different positions on the edge of the welding pool changes with the distance from the center point at the tail of the welding pool. At the weld side boundary, the columnar crystal growth rate (R) tends to be 0. At the center point's midpoint at the welding pool's tail, the columnar crystal growth rate (R) tends to the welding speed V.

Another critical variable determining the microstructure structural characteristics of the weld is the cooling rate (C_R). The weld energy and base metal thickness are the main factors determining the weld cooling rate. Considering the thickness and size of the base metal used in this paper, the cooling rate of the weld can be calculated by Equation (12) [26]:

$$C_R = \frac{2\pi k \rho c l^2 (T_C - T_0)^3}{H_{net}^2} \quad (12)$$

where C_R is the cooling rate ($K \cdot s^{-1}$), k is the thermal conductivity, T_C is the liquid phase line temperature of the molten pool, T_0 is the ambient temperature, H_{net} is the welding wire energy, ρ is the density, c is the specific heat capacity, and l is the plate thickness.

The solid phase temperature gradient (G_S) and liquid phase temperature gradient (G_L) on each side of the solid/liquid interface at the edge of the welding pool play a decisive role in the initial structure of the weld. G_S can be calculated by Equation (13) as follows:

$$G_s = \frac{C_R}{R} \quad (13)$$

G_L plays a critical role in determining the morphology of the solid/liquid interface at the microscopic scale. It is proportional to the energy of the welding line and is strongly affected by the convection inside the welding pool. However, due to the limitation of technical conditions, the temperature gradient cannot be accurately measured, and at the same time, no relevant literature provides reference temperature gradient data. The temperature gradient in the welding pool is a function of material properties, welding process, position in the weld, and heat input, from which the general trend of liquid phase temperature gradient (G_L) with the above factors can be obtained [5]. For the GMAW process, increasing of heat input will increase the welding pool size and reduce the temperature gradient.

The alloy solidification process is mainly affected by "component supercooling," and the solid/liquid solute redistribution process in the welding pool solidification process is the main factor producing component supercooling. The degree of component supercooling at the solid/liquid interface is mainly affected by the degree of solute enrichment and the temperature gradient G_L of the liquid phase of the welding pool. There is always vigorous stirring and convection in the welding pool of the GMAW process, which could significantly affect the liquid temperature gradient G_L near the solid/liquid interface of the welding pool.

Measuring the G_L and component subcooling in the welding process is complicated. Although there is no way to accurately measure the convection intensity, temperature gradient, and solute distribution in the welding pool, it has been shown that the convection enhancement could reduce the temperature gradient G_L [27]. It can be inferred that the convection inside the GMAW welding pool could increase the degree of component undercooling, as shown in Figure 20.

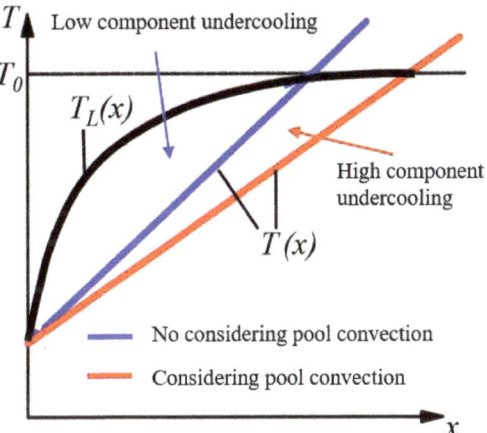

Figure 20. Model of the influence of pool convection on the "component supercooling".

The solute concentration distribution near the solid/liquid interface of the weld pool is shown by $T_L(x)$. In the case of convection and heat exchange of liquid metal in the welding pool, the liquid phase temperature gradient (G_L) decreases, and the actual temperature $T(x)$ near the liquid phase of the solid/liquid interface changes from blue line to red line, which aggravates the degree of component supercooling and then leads to the change of the solidification process.

Figures 21a and 22a show the macroscopic cross-section photos of the weld with different pulse peak currents. The weld penetration depth is 0.7 mm with I_p = 485 A. When I_p = 525 A, the penetration depth increases to 1 mm. This indicates that higher I_p increases weld penetration depth. Figures 23a and 24a show the macroscopic cross-section photos of a single pulse process weld with different pulse waveform shapes. The penetration depth of the trapezoidal pulse weld is 1.56 mm, and of the exponential pulse weld is 1.4 mm. Under the squeezing and pushing action of the pulse peak current, the overheated molten droplet metal directly enters the welding pool, melting the bottom base metal and the increase of the depth of the pool. The weld depth of the single pulse welding process is affected by the droplets' speed, frequency, and temperature. Murray and Scotti et al. [10] analyzed the factors affecting the weld depth of GMAW from the perspective of heat and mass transfer, and the results showed that the heat of the droplets was the main factor affecting the weld depth. The study of Hosh et al. [28] showed that the temperature of the melting drop in the single pulse process was positively correlated with the pulse peak current. As mentioned in Section 3.1, increasing I_p significantly increases the melting velocity. Although it could lead to a decrease in pulse frequency, the equivalent impact force of the melting drop still rises slowly. Under the same welding current, the increase of I_p will simultaneously lead to the increase of the equivalent impact force of the melting drop and the increase of the temperature of the melting drop and then increase the weld penetration depth. With the same I_p, no research has been found on the influence of pulse waveform shape on the temperature of the droplet. However, as described in Section 3.1.2, the equivalent impact force of the trapezoidal pulse on the droplet is higher than that of the exponential pulse, which leads to the increase of the former penetration depth.

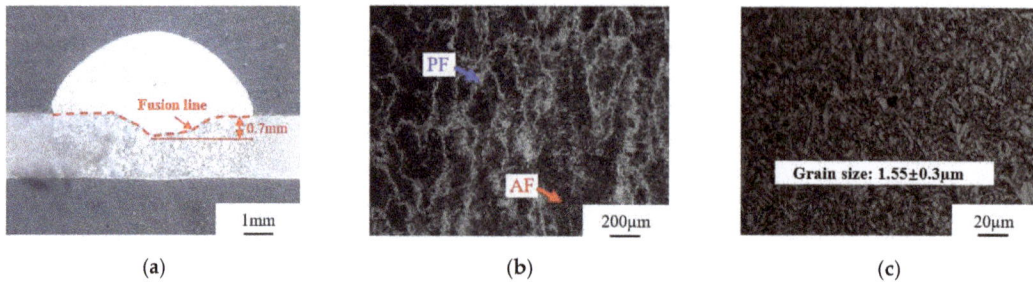

Figure 21. Morphology and structure of weld (I = 60 A, I_p = 485 A): (**a**) macroscopic morphology of weld; (**b**) microstructure at the center of the weld; (**c**) high magnification image of the AF phase.

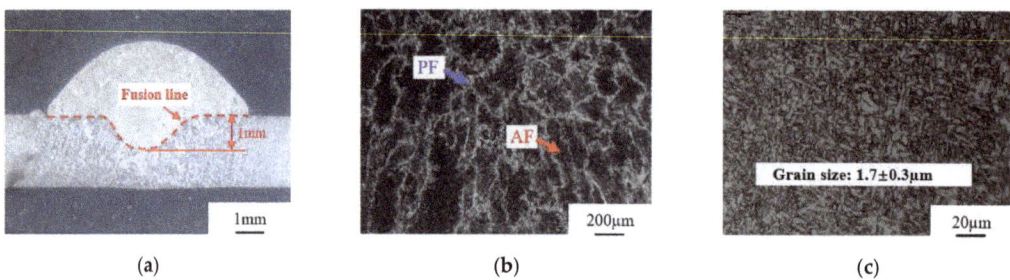

Figure 22. Morphology and structure of weld (I = 60 A, I_p = 525 A): (**a**) macroscopic morphology of weld; (**b**) microstructure at the center of the weld; (**c**) high magnification image of the AF phase.

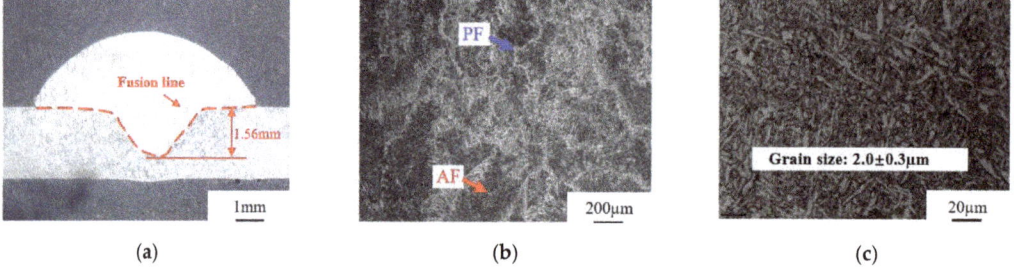

Figure 23. Morphology and structure of weld (I = 100 A, exponential pulse waveform): (**a**) macroscopic morphology of weld; (**b**) microstructure at the center of the weld; (**c**) high magnification image of the AF phase.

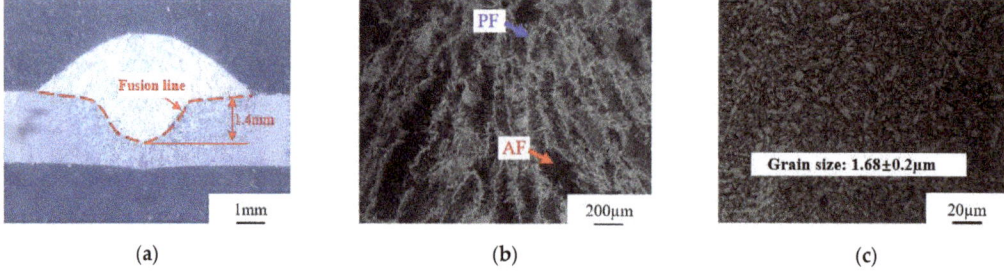

Figure 24. Morphology and structure of weld (I = 100 A, trapezoid pulse waveform): (**a**) macroscopic morphology of weld; (**b**) microstructure at the center of the weld; (**c**) high magnification image of the AF phase.

The peak current and shape of the pulse waveform also affected the weld structure. The weld structures with different I_ps are shown in Figures 21b and 22b, and the weld structures with different pulse waveform shapes are shown in Figures 23b and 24b. The welded structure mainly comprises pro-eutectoid ferrite (PF) and acicular ferrite (AF). Figures 21b and 22b show that the degree of intersection between adjacent PF dendrites in the welded tissue with I_p = 525 A is higher than that of I_p = 485 A. Figure 24b shows the exponential pulse weld tissue, in which the PF dendrites at the center show the same growth direction and are columnar dendrites. However, the PF phase in the trapezoidal pulse weld tissue shows a typical equiaxed dendrite morphology, as shown in Figure 23b. Figures 21c, 22c, 23c and 24 shows the tissue's metallographic photos of the AF phase. The average grain sizes of AF grain were calculated using the intercept method (as per ASTME112-10). It can be found that the pulse waveform parameters have no significant effect on the microstructure of the heat-affected zone with the same welding current, but the grain size of the AF phase is different. This paper measured the proportion of the AF and PF phases in the welded tissue under different pulse waveform parameters, as shown in Figure 25. The results showed that the proportion of the PF phase in the welded tissue with I_p = 525 A was higher than that with I_p = 485 A, and the proportion of the PF phase in the welded tissue with a trapeziform pulse was higher than that with an exponential pulse. At the same time, this section calculates the weld's cooling rate, growth rate, and temperature gradient under different pulse waveform parameters, and the results are shown in Table 2. The cooling rate, growth rate, and temperature gradient of the weld under different pulse waveforms are all consistent with the same welding parameters, and the difference in the dynamic behavior of the weld pool caused by pulse waveforms is the main factor affecting the weld structure.

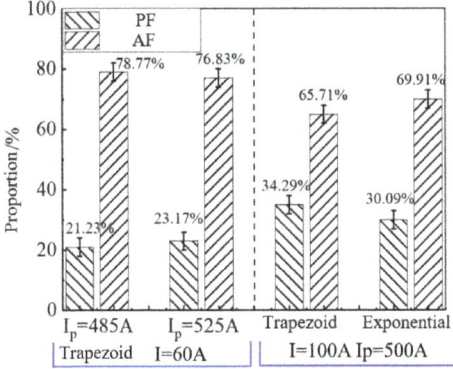

Figure 25. The volume proportion of PF phase and AF phase in weld microstructure.

Table 2. The welding parameters used in this article.

Waveform	I/A	V/cm·min^{-1}	I_p/A	H_{net}/J mm^{-1}	C_R/K s^{-1}	R/mm s^{-1}	G_s/K·mm^{-1}
Trapezoidal type	60	25	485	312.3	150.7	4.1	37.2
	60	25	525	310.9	152.1	4.1	36.5
	100	38.5	500	345.6	128.8	6.4	20.1
Exponential type	100	38.5	500	338.1	129.8	6.4	20.3

The convection intensity of the welding pool in the P-GMAW process is positively correlated with the droplet impact. The convection intensifies the subcooling degree during the solidification process and then changes the growth orientation of the primary austenite grain of the weld and reduces the cooling rate of the solidified weld area. Under different pulse waveform parameters, the droplet impact increase could lead to the intersection degree of adjacent PF dendrites in the weld microstructure and the transformation of the

PF phase from columnar to equiaxial dendrites. At the same time, it also slows down the cooling rate of the solidified weld microstructure below the tail of the weld pool, resulting in a large amount of PF phase precipitation and an increase in AF phase grain size.

4. Conclusions

(1) The arc pressure is positively correlated with the I_p. The extrusion of electromagnetic contraction force affects the droplet transition process and the impact of charged particles in the arc space. The droplet velocity and momentum are positively correlated with the peak pulse current.

(2) The change of pulse waveform shape does not affect the peak pressure of the arc with the same pulse waveform parameters. Compared with the exponential pulse, the trapezoidal current pulse droplet has a shorter time to detach from the welding wire. The droplet is faster and has apparent deformation, leading to a more significant impact on the welding pool with the same parameter conditions.

(3) The welding pool's convection intensity and weld depth are positively correlated with the impact of the arc and droplet. The liquid convection reduces the temperature gradient of the welding pool, intensifies the component supercooling, and significantly changes the growth orientation of the primary austenite phase. The stronger the impact, the higher the proportion of PF phase in the weld microstructure, the larger the grain size of the PF phase and AF phase, and the intersection degree of adjacent PF phase dendrites increases.

Author Contributions: Conceptualization, J.H. and T.C.; methodology, J.H. and T.C.; software, J.H. and T.C.; validation, J.H. and T.C.; formal analysis, J.H. and T.C.; investigation, J.H. and T.C.; resources, J.H. and T.C.; data curation, T.C.; writing—original draft preparation, J.H.; writing—review and editing, J.H. and T.C.; visualization, J.H., T.C., D.H. and T.X.; supervision, J.H.; project administration, J.H.; funding acquisition, J.H. All authors have read and agreed to the published version of the manuscript.

Funding: This research was funded by the Ministry of Science and Technology Innovation Method Work Special Project: The Construction and Demonstration of Multi-level and Multi-mode College Innovation Method Talent Cultivation System, grant number 2017IM030100 and Education reform in Jiangsu Province: Exploration and practice of the "three-step" training model for top-notch talents through double innovation in higher vocational colleges, grant number 2017JSJG475.

Data Availability Statement: Data is available on request from the authors.

Conflicts of Interest: The authors declare no conflict of interest.

References

1. Richards, R.W.; Jones, R.D.; Clements, P.D.; Clarke, H. Metallurgy of continuous hot dip aluminizing. *Int. Mater. Rev.* **2013**, *39*, 13–19.
2. Kah, P.; Suoranta, R.; Martikainen, J. Advanced gas metal arc welding processes. *Int. J. Adv. Manuf. Technol.* **2013**, *67*, 655–674. [CrossRef]
3. Fan, H.G.; Kovacevic, R. A unified model of transport phenomena in gas metal arc welding including electrode, arc plasma and molten pool. *J. Phys. D Appl. Phys.* **2004**, *37*, 2531. [CrossRef]
4. Min, H.C.; Fa Rson, D.F. Understanding Bead Hump Formation in Gas Metal Arc Welding Using a Numerical Simulation. *Metall. Mater. Trans. B* **2007**, *38*, 305–319.
5. David, S.A.; Vitek, J.M. Correlation between solidification parameters and weld microstructures. *Metall. Rev.* **2013**, *34*, 213–245. [CrossRef]
6. Guo, J.; Zhou, Y.; Liu, C.; Wu, Q.; Chen, X.; Lu, J. Wire Arc Additive Manufacturing of AZ31 Magnesium Alloy: Grain Refinement by Adjusting Pulse Frequency. *Materials* **2016**, *9*, 823. [CrossRef]
7. Jolanta, M.; Tomasz, P. The research of technological and environmental conditions during low-energetic gas-shielded metal arc welding of aluminium alloys. *Weld. Int.* **2013**, *5*, 338–344.
8. Liu, A.; Tang, X.; Lu, F. Study on welding process and prosperities of AA5754 Al-alloy welded by double pulsed gas metal arc welding. *Mater. Des.* **2013**, *50*, 149–155. [CrossRef]
9. Scotti, A.; Rodrigues, C. Determination of momentum as a mean of quantifying the mechanical energy delivered by droplets during MIG/MAG welding. *Eur. Phys. J. Appl. Phys.* **2009**, *45*, 11201. [CrossRef]
10. Depth of penetration in gas metal arc welding. *Sci. Technol. Weld. Join.* **1999**, *4*, 111–118.

11. Hussain, H.M.; Ghosh, P.K.; Gupta, P.C.; Potluri, N.B. Fatigue crack growth properties of pulse current multipass mig-weld of al-zn-mg alloy. *Trans. Indian Inst. Met.* **1997**, *50*, 275–285.
12. Ghosh, P.K.; Yongyuth, P.; Gupta, P.C. Two-Dimensional Spatial Geometric Solution for Estimating the Macroconstituents Affecting the Mechanical Properties of Mulitpass C-Mn Steel SAW Deposits. *Isij Int.* **2007**, *35*, 63–70. [CrossRef]
13. Cong, B.Q.; Yang, M.X.; Qi, B.J.; Li, W. Effects of pulse parameters on arc characteristics and weld penetration in hybrid pulse VP-GTAW of aluminum alloy. *China Weld.* **2010**, *19*, 68–73.
14. Pal, K.; Pal, S.K. Effect of pulse parameters on weld quality. *J. Mater. Eng. Perform.* **2011**, *20*, 918–931. [CrossRef]
15. Greene, W.J. An analysis of transfer in gas-shielded welding arcs. *Trans. Am. Inst. Electr. Eng. Part II Appl. Ind.* **1960**, *79*, 194–203.
16. Arif, N.; Lee, J.H.; Yoo, C.D. Modelling of globular transfer considering momentum flux in GMAW. *J. Phys. D Appl. Phys.* **2008**, *41*, 195503. [CrossRef]
17. Lancaster, J.F. The physics of fusion welding. Part 1: The electric arc in welding. *IEE Proc. B. Electr. Power Appl.* **2008**, *134*, 233–254. [CrossRef]
18. Yudodibroto, M.; Hermans, M. Observations on Droplet and Arc Behavior during Pulsed GMAW. *Weld. World* **2009**, *53*, 171–180. [CrossRef]
19. Wu, C.S.; Chen, M.A.; Lu, Y.F. Effect of current waveforms on metal transfer in pulsed gas metal arc welding. *Meas. Sci. Technol.* **2005**, *16*, 2459. [CrossRef]
20. Hertel, M.; Spille-Kohoff, A.; Füssel, U.; Schnick, M. Numerical simulation of droplet detachment in pulsed gas–metal arc welding including the influence of metal vapour. *J. Phys. D Appl. Phys.* **2013**, *46*, 224003. [CrossRef]
21. Hu, J.; Tsai, H.L. Heat and mass transfer in gas metal arc welding. Part I: The arc. *Int. J. Heat Mass Transf.* **2007**, *50*, 833–846. [CrossRef]
22. Hu, J.; Tsai, H.L. Heat and mass transfer in gas metal arc welding. Part II: The metal. *Int. J. Heat Mass Transf.* **2007**, *50*, 808–820. [CrossRef]
23. Haidar, J. The dynamic effects of metal vapour in gas metal arc welding. *J. Phys. D Appl. Phys.* **2010**, *43*, 165204. [CrossRef]
24. Rouffet, M.E.; Wendt, M.; Goett, G.; Kozakov, R.; Schoepp, H.; Weltmann, K.D.; Uhrlandt, D. Spectroscopic investigation of the high-current phase of a pulsed GMAW process. *J. Phys. D Appl. Phys.* **2011**, *43*, 434003. [CrossRef]
25. Ahsan, M.R.; Cheepu, M.; Ashiri, R.; Kim, T.H.; Jeong, C.; Park, Y.D. Mechanisms of weld pool flow and slag formation location in cold metal transfer (CMT) gas metal arc welding (GMAW). *Weld. World* **2017**, *61*, 1275–1285. [CrossRef]
26. Sen, M.; Mukherjee, M.; Singh, S.K.; Pal, T.K. Effect of Double-Pulsed Gas Metal Arc Welding (DP-GMAW) Process Variables on Microstructural Constituents and Hardness of Low Carbon Steel Weld Deposits. *J. Manuf. Process.* **2017**, *31*, 424–439. [CrossRef]
27. Yuan, T.; Luo, Z.; Kou, S. Grain refining of magnesium welds by arc oscillation. *Acta Mater.* **2016**, *116*, 166–176. [CrossRef]
28. Ghosh, P.K.; Dorn, L.; Hübner, M.; Goyal, V.K. Arc characteristics and behaviour of metal transfer in pulsed current GMA welding of aluminium alloy. *J. Mater. Process. Technol.* **2007**, *194*, 163–175. [CrossRef]

Disclaimer/Publisher's Note: The statements, opinions and data contained in all publications are solely those of the individual author(s) and contributor(s) and not of MDPI and/or the editor(s). MDPI and/or the editor(s) disclaim responsibility for any injury to people or property resulting from any ideas, methods, instructions or products referred to in the content.

Article

Effect of Ce Content on the Microstructure and Mechanical Properties of Al-Cu-Li Alloy

Xianxian Ding [1], Yalin Lu [1,2,*], Jian Wang [1], Xingcheng Li [2] and Dongshuai Zhou [2]

[1] School of Materials Engineering, Jiangsu University of Technology, Changzhou 213000, China
[2] Key Construction Laboratory of Green Forming and Equipment from Jiangsu Province, Changzhou 213000, China
* Correspondence: luyalin@163.com

Abstract: In this work, the effects of Ce content (0.1%, 0.2%, and 0.3 wt%) on the microstructures and mechanical properties of Al-Cu-Li alloys were investigated. The results show that the grains of Al-Cu-Li alloy are refined by adding Ce element. When Ce content is less than 0.1%, the ultimate strength of the alloy increases with the increase of Ce content. However, the ultimate strength of the alloy decreases when Ce content is above 0.1%. For Al-Cu-Li alloy with different Ce content, the aging precipitation of T_1 and θ' phases play the main strengthening role. When Ce content increases from 0.1 to 0.3%, dynamic recrystallization is promoted during hot deformation. The recrystallization of the alloys is inhibited after T6 treatment with the increase of Ce content, which can be attributed to existence of the Al_8Cu_4Ce phases on the grain boundary. This work provides an economical and convenient method for improving the properties of Al-Cu-Li alloys by Ce addition.

Keywords: Al-Cu-Li alloy; RE cerium; microstructure; mechanical properties

1. Introduction

In order to meet the requirement of some components with high-performance and lightweight, aluminum–lithium alloys have been applied in the fields of aerospace and transportation owing to their advantages, such as low density, low fatigue crack growth rate, high specific strength, and good high-temperature and low-temperature properties [1–5]. So far, three generations of aluminum–lithium alloys have been developed. The third-generation aluminum–lithium alloys, including 2195, 2050, 2099, 2198, and 2199, have been successfully used to produce the carrier rocket tank and aircraft skins, a weight reduction of about 10–20%, and a strength increase of about 15–20% [6,7]. Currently, many researchers are devoted to developing fourth-generation aluminum–lithium alloys.

Rare earth micro-alloying is one of the important methods to produce novel aluminum–lithium alloys. Some rare earth elements such as Ce, Sc, and Er could be utilized as the beneficial alloying additions in Al-Cu-Li based alloys to improve their microstructures and mechanical properties [8]. Suresh et al. [9] found that the grain was refined, and precipitation kinetics were enhanced by adding Sc to AA2195 alloy. Yu et al. [10] found that the recrystallization was restrained, and the strength of the alloy increased by adding 0.1 wt% Er to Al-3Li-1Cu-0.1Zr alloy. However, Sc, as an effective rare earth refiner, is limited by its high price [11]. Due to the high proportion of Ce content in rare earth minerals, Ce elements are mostly obtained when refining useful rare earth elements. The effects of Ce element on the microstructure of aluminum alloy are as follows [12,13]. Firstly, the diffusion of the main elements for the alloy is hindered and the coarsening of the primary phase is delayed because of the addition of Ce element. Secondly, the primary AlCuCe phase acts as a nucleating agent and increases the region of compositional supercooling, which finally reduces the distance between the secondary dendrites. Thirdly, τ_1 (Al_8Cu_4Ce) [14–16] phase dispersion occurs during homogenization and thermomechanical processes, which inhibits recrystallization during the subsequent heat treatment. Chaubey et al. [17] found

that the morphology of the precipitates changed from spherical to needle-like and the resistance from dislocation movement increased when the Ce content changed from 0.1% to 0.4% (mass fraction). Meanwhile, the strength of the alloy changed with the increase of Ce content. Yu et al. [18] found that the strength of the alloy increased by adding 0.2% Ce to Al-5.87Cu-1.31Li alloy. Ma et al. [19] found the strength of the alloy decreased when the Ce content of Al-4.24Cu-1.26Li alloy was about 0.11%.

Generally, the properties of aluminum alloys can be improved by means of inhibiting recrystallization. The strength and fracture toughness of aluminum alloy has been greatly improved by inhibiting recrystallization of the alloy, increasing the number of small grains and fibrous microstructures [20–23]. For Al-Cu-Li alloys, recrystallization inhibition can avoid intergranular fracture by reducing lithium segregation at the low angle grain boundary (LAGBs) and then improving the strength of the alloy [24,25]. Yu et al. [26] found that the addition of Ce element inhibited recrystallization of the alloy and improved the yield strength and fracture toughness by adding 0.29wt% Ce to Al-Cu-Li-0.13Zr alloy.

Nowadays, the effect of Ce element on the mechanical property and inhibition recrystallization of aluminum lithium alloy has not been systematically studied. In order to clarify the micro-alloying effect of the Ce element, Al-3.5Cu-1.2Li alloys with different Ce content were fabricated in our work. The mechanical properties and microstructural evolution of the alloy after rolling and heat treatment were investigated in this paper. The feasibility of Ce addition to improve the alloy strength and the addition of optimal Ce addition are discussed in detail. This study can provide an economical and convenient method for performance enhancement.

2. Experimental Materials and Procedures

The experimental materials were Al-3.5Cu-1.2Li-0.5Mg-0.3Mn-0.3Zn-0.11Zr (wt.%) with different Ce contents, as shown in Table 1. The alloys in this paper were called AL1-AL4, respectively, according to the Ce contents. In order to fabricate the alloys, pure Al, Li, Zn, Mg, and master alloys of Al-Cu, Al-Mn, and Al-Zr were melted in a high-purity graphite crucible inside a vacuum high-frequency induction furnace and were poured into the pre-heated mold. Then, the experimental ingots of AL1-AL4 were obtained.

Table 1. Chemical composition of experimental alloys (wt.%).

Alloy	Cu	Li	Mg	Mn	Zn	Zr	Ce	Al
AL1	3.5	1.2	0.5	0.3	0.3	0.11	–	Bal.
AL2	3.5	1.2	0.5	0.3	0.3	0.11	0.1	Bal.
AL3	3.5	1.2	0.5	0.3	0.3	0.11	0.2	Bal.
AL4	3.5	1.2	0.5	0.3	0.3	0.11	0.3	Bal.

The experimental plates of AL1-AL4 alloy with dimensions of 110 mm × 40 mm × 20 mm were taken from the ingot by Electrospark Wire-Electrode Cutting (BMG Co., Ltd., Suzhou, China). The specimens were heat treated at a temperature of 470 °C for 8 h and a temperature of 520 °C for 24 h, and then cooled in air. The specimens were rolled at a temperature of 420 °C until the thickness of the specimens was 4 mm and the deformation extent of the specimen was about 80%. After the rolling process, the specimens were solution-treated at a temperature of 500 °C for 2 h and quenched in water within 3 s. Then, the specimens were age-treated at a temperature of 175 °C for 24 h.

In order to carry out the microstructure observation and mechanical property measurement, the specimens were machined along the rolling direction (RD), normal direction (ND), and transverse direction (TD). The tensile properties of the alloys after heat treatment were measured with a CMT-5105 testing machine (WANCE Co., Ltd., Shenzhen, China) at rate of 2 mm/min. The results were obtained from the average value of five tensile specimens in the same condition.

The microstructures of the specimens were characterized using an optical microscope (OM, Zeiss, Primotech, Berlin, Germany), the X-ray diffraction (XRD, Bruker, Karlsruhe, Germany), the electron back-scatter diffraction (EBSD, Carl Zeiss, Oberkoichen, Germany), and a JEOL-2100F transmission electron microscope (TEM, FEI Company, Hillsboro, OR, USA). The samples for OM observation were mechanically ground, then etched in standard Keller's reagent (1 mL HF + 1.5 mL HCl + 2.5 mL HNO_3 + 95 mL H_2O). The samples for EBSD observation were firstly mechanically ground and then electrochemically polished in the mixed 10% $HClO_4$ and 90% C_2H_5OH solution at a temperature of −20 °C. In order to perform the TEM analysis, the specimens were mechanically thinned to thicknesses of about 60 μm, punched into foils of 3 mm in diameter, and electro-polished in 30% HNO_3 and 70% CH_3OH solution below −30 °C. The analysis was performed at an acceleration voltage of 200 kV. Positions and dimensions of the specimen for microstructure observation and tensile test are shown in Figure 1.

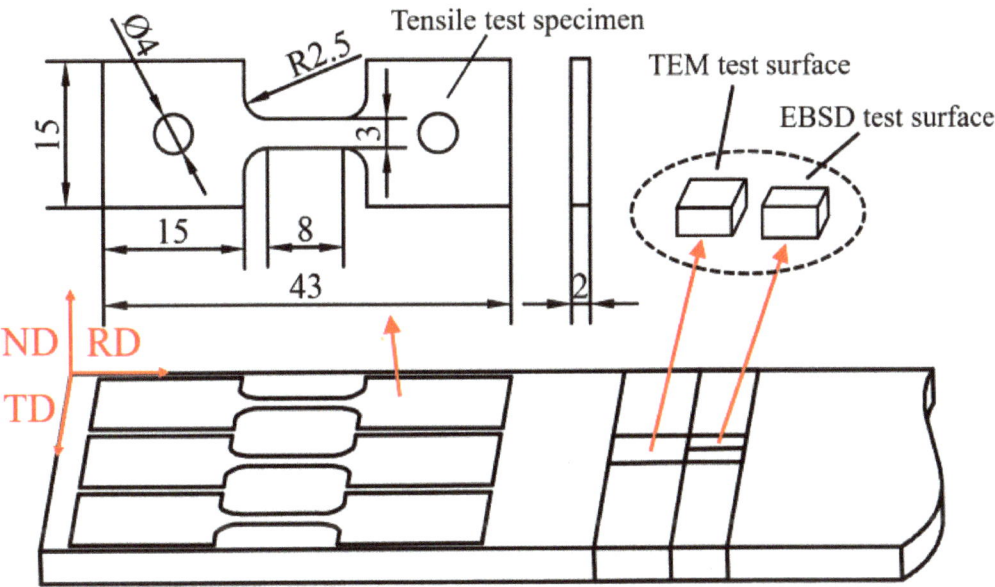

Figure 1. Positions and dimensions of the specimens for microstructure observation and tensile test.

3. Results and Discussion

3.1. Microstructures of As-Cast Alloys

Figure 2 illustrates the optical micrographs of as-cast AL1, AL2, AL3, and AL4 alloys. It can be seen that the α-Al dendrites were refined with the addition of Ce. As shown in Figure 2a, the coarse dendrite microstructure can be seen in the as-cast microstructure of the AL1 alloy. For AL2 alloy, the dendrites are still relatively coarse, but the dendrites were refined compared to AL1 alloy, as shown in Figure 2b. With the increase of Ce content, the dendrites were further refined, and the secondary dendrites arm spacing was reduced at the same time, as shown in Figure 2c,d. The coarse dendrites were refined, and the secondary dendrites arm spacing was reduced with the increase of Ce content.

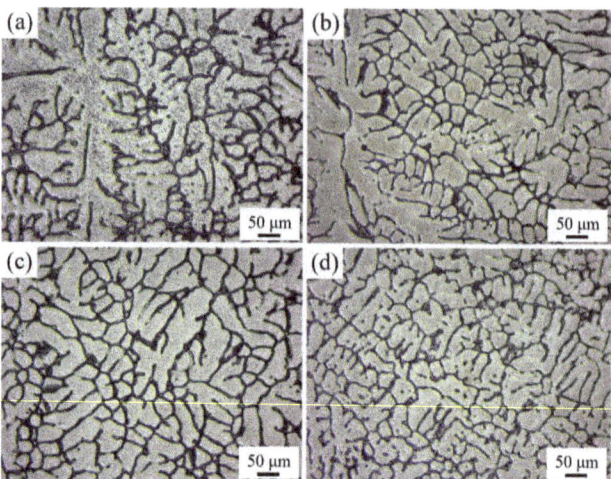

Figure 2. Optical micrographs of the as-cast alloys with different Ce contents (**a**) AL1; (**b**) AL2; (**c**) L3; and (**d**) AL4.

Optical micrographs of AL1-AL4 alloys after homogenization treatment are shown in Figure 3. It can be found that the dendritic structures almost disappear after homogenization treatment. Most of the secondary phases in as-cast alloy have been dissolved into α-Al matrix.

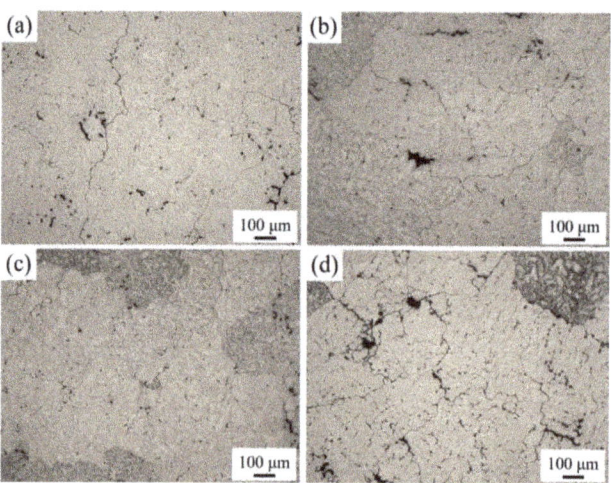

Figure 3. Optical micrographs of homogenized alloys with different Ce contents (**a**) AL1; (**b**) AL2; (**c**) AL3; and (**d**) AL4.

3.2. Microstructures of the Alloys after Hot Rolling

To investigate the effect of Ce element after rolling process on the microstructures of AL1-AL4 alloys, EBSD analysis was used in our work. Figure 4 shows the inverse pole figures and orientation maps of four alloys after 80% rolling reduction. Low angle grain boundaries (LAGB) are marked with white lines, which are defined as misorientation $3° < θ < 15°$. High angle grain boundaries (LAGB) are marked with black lines, which are defined as misorientation $θ > 15°$. Different crystal orientations are represented by different colors.

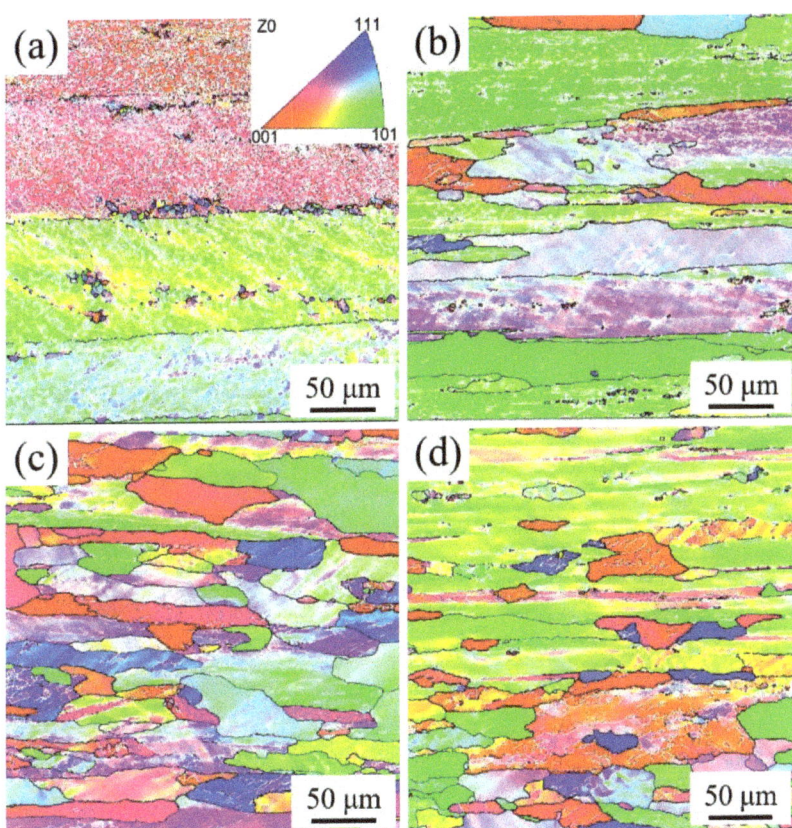

Figure 4. Orientation maps and inverse pole figure (**a**) AL1; (**b**) AL2; (**c**) AL3; (**d**) AL4.

As shown in Figure 4a, the grains of AL1 alloy are elongated significantly after hot rolling with fibrous shape along the rolling direction, exhibiting a large aspect ratio. There are a large number of LAGB in the fibrous shape grains, which indicates the presence of a few recrystallization grains during hot rolling. When the Ce addition is 0.1%, the microstructures keep elongated grains and the number of grain increased, as shown in Figure 4b. Meanwhile, it can be seen that some recrystallized structures appear, and the degree of recrystallization increases compared with AL1 alloy. With the increase of Ce content to 0.2%, the number of coarse elongated grains of the alloy decreased. The grain aspect ratios reduce and grain shape tends to spheroidize, as shown in Figure 4c. When the Ce content increases to 0.3%, the grain size of the AL4 alloy changes slightly, as shown in Figure 4d. However, the low angle grain boundary is significantly lower than that in Figure 4c.

Figure 5 illustrates the microstructures of AL1-AL4 alloys after hot rolling. The deformation grains, sub-structured grains (recovered with sub-grains), and recrystallized grains of four alloys are shown in Figure 5 in three colors (red, yellow, and blue, respectively). According to average misorientation angles, these grains are classified as recrystallized grains when their average misorientation angle is above 15°, and as sub-structured grains if their average misorientation angle is between 3° and 15°. The grains are defined as deformation grains when their average misorientation angle is below 3°.

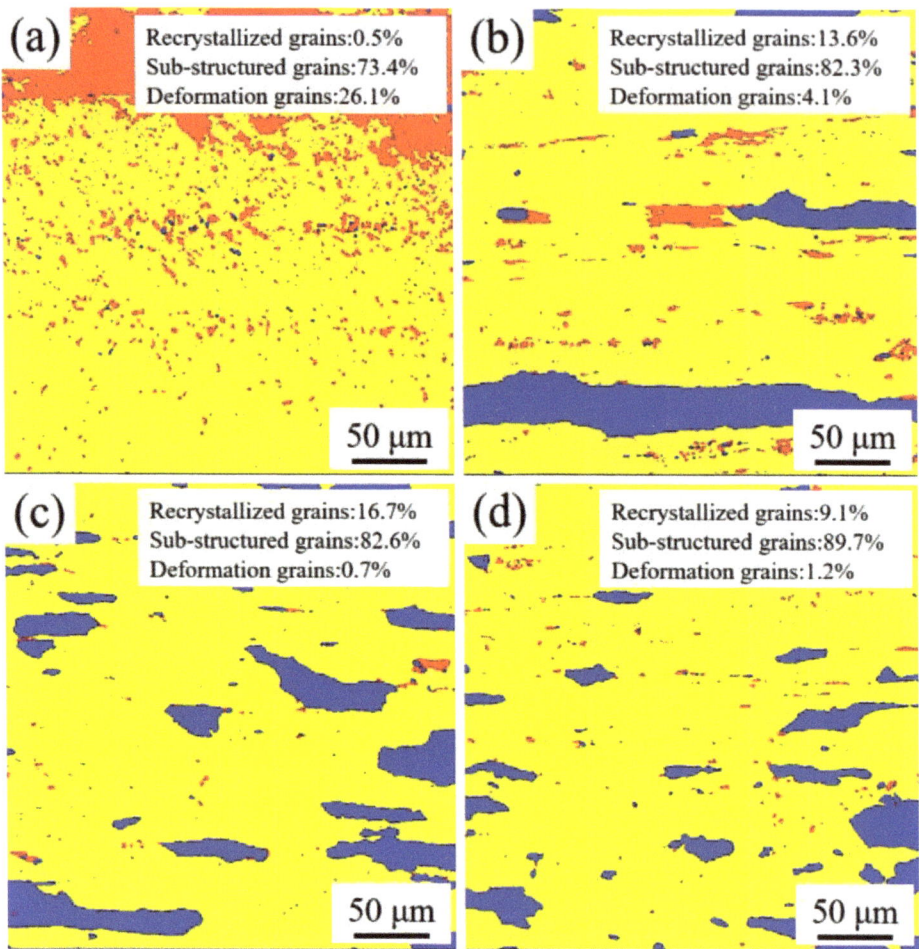

Figure 5. Microstructures of the alloys after rolling process (**a**) AL1; (**b**) AL2; (**c**) AL3; (**d**) AL4.

The dynamic recrystallization degree of AL1 alloy is about 0.5% after the rolling process, as shown in Figure 5a. The deformation structure of the alloy changes with the increase of Ce content, as shown in Figure 5b. An incomplete recrystallization consisting of substructure grains, recrystallization grains, and deformation grains appears when the Ce content is 0.1%. The degree of recrystallization of AL2 alloy increases from 0.5% to 13.6%. When the Ce content increases to 0.2%, the degree of recrystallization of the alloy increases slightly, as shown in Figure 5c. However, when the Ce content reaches 0.3%, the degree of recrystallization of AL4 alloy decreases from 16.7% to 9.1%.

Figure 6 illustrates the misorientation distributions and the fractions of HAGBs of AL1-AL4 alloys. The misorientation below 3° has been removed from Figure 6. It can be found that the fraction of LAGBs is much higher than that of HAGBs. For AL1 alloy, the average misorientation angle is 2.5°, and the fraction of HAGBs is 2.4% as shown in Figure 6a. When the Ce content increases from 0.1%, 0.2%, to 0.3%, the fractions of HAGBs are 5.6%, 10.3%, and 8.2%, respectively. Meanwhile, the average misorientation angles are 3.0°, 4.6°, and 4.0°, respectively, as illustrated in Figure 6b–d.

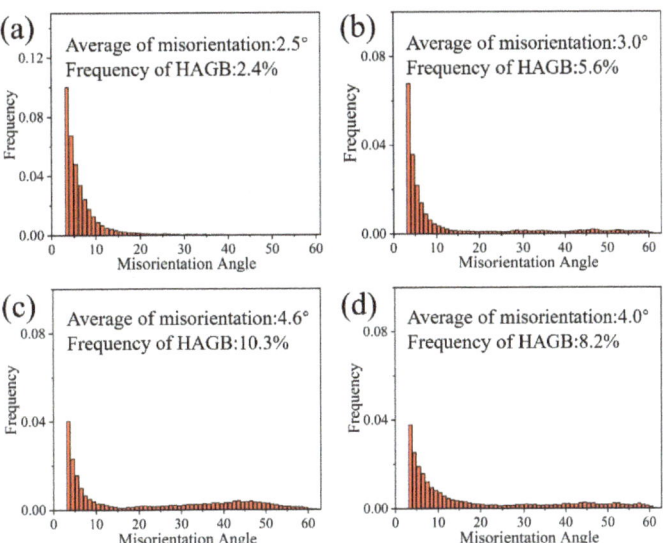

Figure 6. Distribution of misorientation angles of the alloys (**a**) AL1; (**b**) AL2; (**c**) AL3; (**d**) AL4.

Figure 7 illustrates the TEM images of AL1 and AL2 alloys after the rolling process. AL1 alloy exhibits typical deformed microstructure with a large number of dislocation structures after hot rolling. Compared with AL1 alloy, there are more dislocation structures in the AL2 alloy, as shown in Figure 7b. From Figure 7c, it can be seen that the dislocations entangle together and form a dislocation wall and sub-grains. The boundary of sub-grains is shown in Figure 7d.

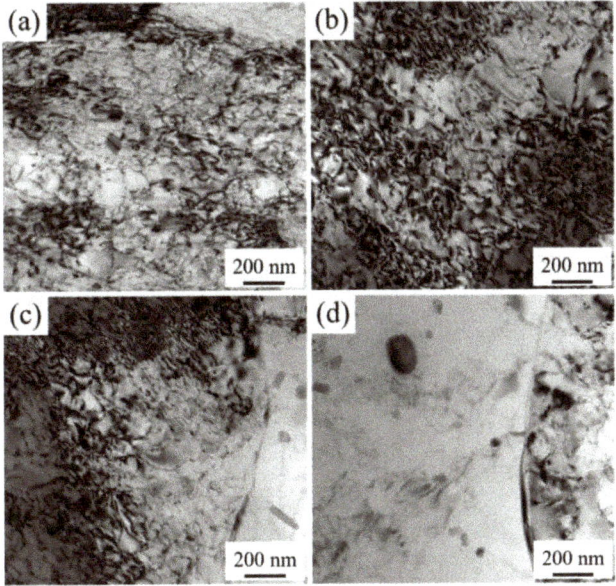

Figure 7. TEM images of the alloys after hot rolling (**a**) AL1; (**b**–**d**) AL2.

3.3. Microstructure and the Mechanical Properties of the Alloys after T6 Treatment

To obtain the microstructure morphologies and the extent of recrystallization of the alloys, EBSD analysis was carried out for the alloys with different Ce contents after T6 treatment. Figure 8 shows the inverse pole figure and orientation maps of AL1- AL4 alloys after T6 treatment. For AL1 alloy without Ce element, there are many elongated grains, and the average grain size is 51 μm, as shown in Figure 8a. From Figure 8b,c and d, the microstructure of AL2, AL3, and AL4 alloys changes gradually. The recrystallization grains occur at the triangle boundary, the grains size decrease, and the number increases with the increase of Ce contents. The average grain sizes are 44 μm, 43 μm, and 39 μm, respectively. As can be seen, the grains of the alloys are refined and spheroridized with the increase of Ce content after T6 treatment.

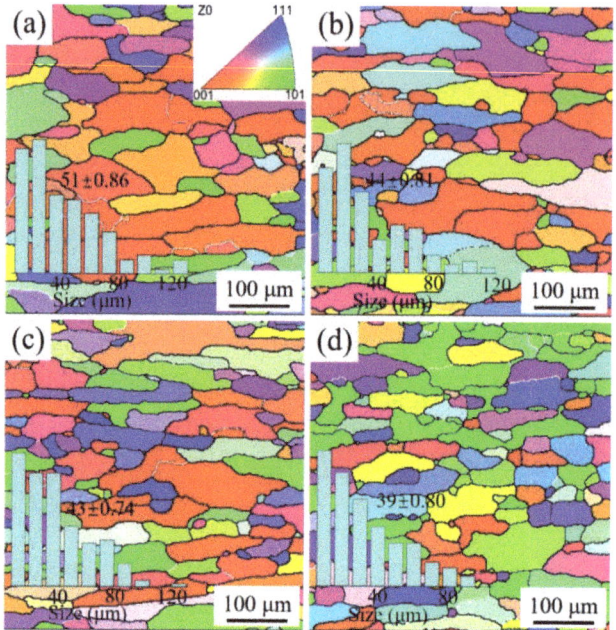

Figure 8. Orientation maps and inverse pole figure of the alloys after T6 treatment (**a**) AL1; (**b**) AL2; (**c**) AL3; (**d**) AL4.

Figure 9 illustrates the recrystallized microstructure of AL1-AL4 alloys after T6 treatment. Similar to Figure 5, the deformation grains, sub-structured grains (recovered with sub-grains), and recrystallized grains of the four alloys are shown in three colors (red, yellow, and blue, respectively). From Figure 9, the degree of recrystallization of the alloys decreases significantly with the increase of Ce content. For AL1 alloy, the fraction of recrystallization is relatively low, and the fractions of recrystallization grains and sub-structured grains are 82.07%, and 17.92%, respectively, as shown in Figure 9a. It can be observed in Figure 9b that the fractions of recrystallization grains and sub-structured grains of AL2 alloy are 96.96% and 3.03%, respectively. It indicates that the recrystallization grains increase by adding Ce content to Al-Cu-Li alloy, which is helpful to refine microstructure of the alloy. However, when Ce content reaches 0.2% and 0.3%, the fractions of recrystallization are 76.24% and 45.47%, respectively, as illustrated in Figure 9c,d. Therefore, when Ce content is above critical value, the degree of recrystallization decreases with the increase of Ce content.

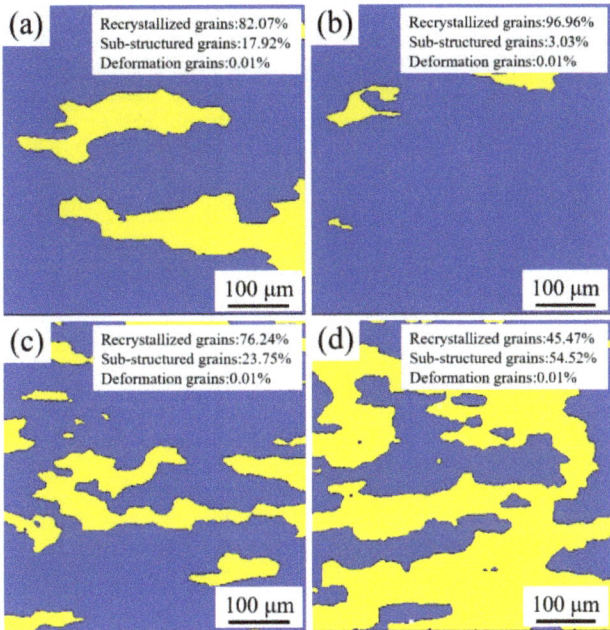

Figure 9. Recrystallized maps of the alloys after T6 treatment (**a**) AL1; (**b**) AL2; (**c**) AL3; (**d**) AL4.

Figure 10 shows the misorientation distributions and the fraction of HAGBs of AL1-AL4 alloys after T6 treatment. The misorientation below 3° has been removed from Figure 10. Compared with Figure 6, it is clear that the LAGBs have transformed into HAGBs. The fraction of HAGBs for AL1 alloy is 30.5%, and the average misorientation angle is 12.5°, as shown in Figure 10a. When the Ce content of the alloy changes from 0.1%, 0.2%, to 0.3%, the fractions of HAGBs are 42.4%, 24.9%, and 28.5%, respectively. Meanwhile, the average misorientation angles are 18.1°, 10.4°, and 11.5°, respectively, as shown in Figure 10b–d.

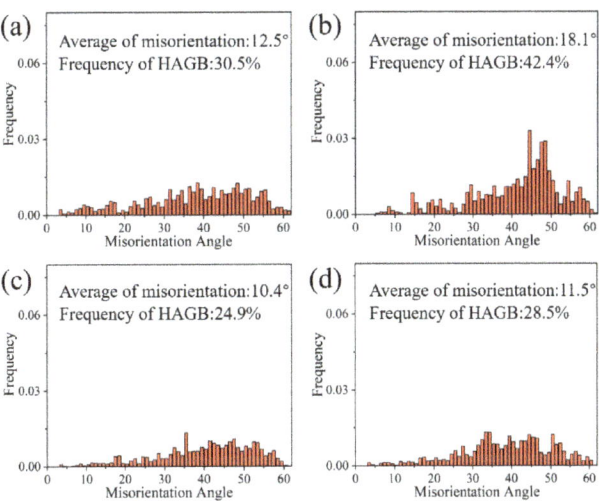

Figure 10. Distribution of misorientation angles of the alloys after T6 treatment (**a**) AL1; (**b**) AL2; (**c**) AL3; (**d**) AL4.

To clarify the effects of Ce enrichment particles in Al-Cu-Li alloys on inhibition of recrystallization. The Ce enrichment phases were investigated by XRD, and the results are shown in Figure 11. The XRD patterns clearly show the presence of Al_8Cu_4Ce phases in AL2-AL4 alloys after T6 treatment. The improved recrystallization resistance achieved by Ce addition can be attributed to the formation of these Al_8Cu_4Ce phases in Al-Cu-Li alloy [26]. These Al_8Cu_4Ce particles might exert a Zener pressure on the grain boundary, which would retard their migration and inhibit recrystallization [27].

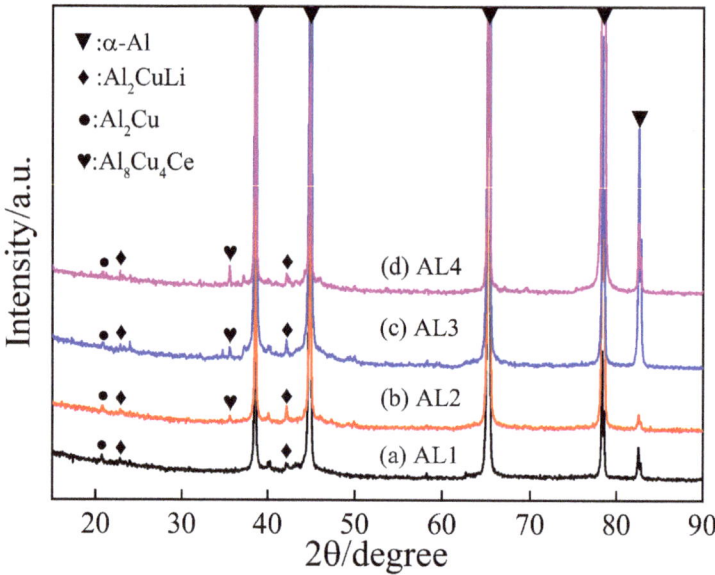

Figure 11. XRD patterns of the alloys after T6 treatment (**a**) AL1; (**b**) AL2; (**c**) AL3; (**d**) AL4.

Figure 12 shows the selected area electron diffraction (SAED) spectra, TEM bright-field and dark-field images of AL2 alloy with 0.1% Ce and the AL1 alloy without Ce element. As shown in Figure 12a, it can be found that there are bright θ' phases in the SAED spectrum along the direction of <100> of AL2 alloy, and the θ' phases distributed uniformity. It indicates that a large number of θ' phases have precipitated in AL2 alloy. Figure 12b shows the θ' phases appear in the TEM bright field image along the direction of <100> of AL1 alloy. However, the number and density of θ' phases are relatively fewer compared with AL2 alloy. Bright T_1 phases can be observed in the SAED spectrum along the direction of <112> of AL2 alloy, and there are T_1 phases parallel to each other in dark-field, as shown in Figure 12c. This indicates that a large number of T_1 phases have precipitated in AL2 alloy. From Figure 12d, T_1 phases appear in the TEM dark field image along the direction of <112> of AL1 alloy. The aging precipitated phases in AL1 alloy are T_1 and θ'. Compared with AL1 alloy, T_1 and θ' phases in AL2 alloy are distributed uniformly.

The mechanical properties of AL1-AL4 alloys after T6 treatment are illustrated in Figure 13. It can be found that the ultimate tensile strength increases with the increase of Ce content when the Ce content is less than 0.1%. However, the ultimate tensile strength of the alloy decreases when the Ce content increase to 0.2% and 0.3%. The yield strength exhibits the same trend, while the change of elongation is not obvious, as shown in Figure 13b. Compared with AL1 alloy, when Ce content is 0.1%, the ultimate tensile strength of AL2 alloyincreasese from 484 MPa to 514 MPa.

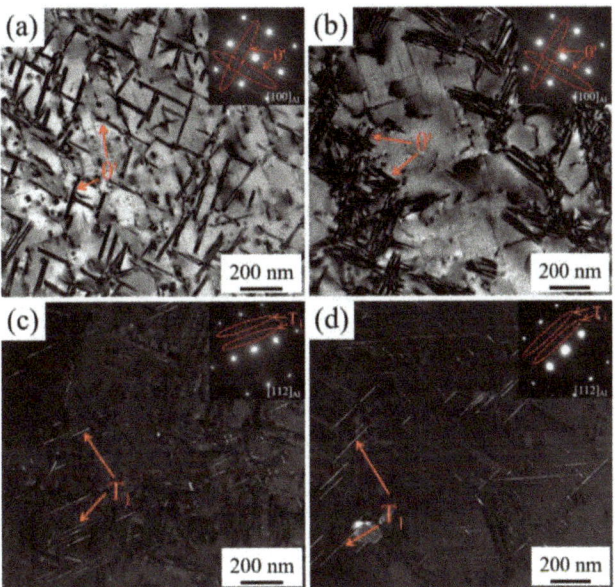

Figure 12. SAED patterns and dark field (DF) and bright field (BF) TEM images of AL2 (**a**), (**c**), and AL1 (**b**), (**d**) alloys (**a**), (**b**) θ′ precipitates (direction parallels to <100>Al); (**c**), (**d**) T_1 precipitates (direction parallels to <112>Al).

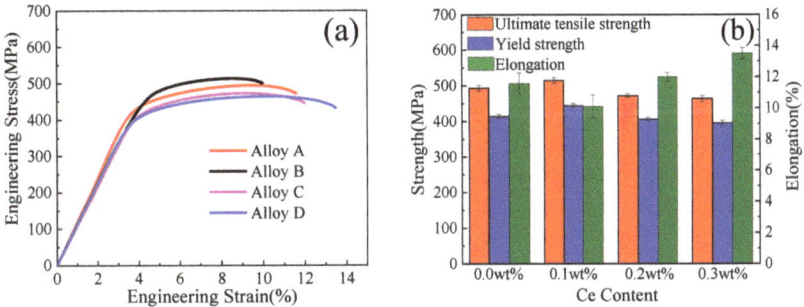

Figure 13. Tensile test of the alloys after T6 treatments. (**a**) Engineering stress–strain curves; (**b**) Tensile properties curves.

3.4. Discussion of Effecting Mechanism

The above results show that the microstructure evolution and the tensile properties of the alloys are mainly related to the Ce content, the grain size and number, and the distribution of the main precipitated phases. Some studies [28,29] show that the Al-Cu-Li alloy is strengthened mainly by T_1 and θ′ phases precipitated from the solid solution matrix during the aging process. The strengthening effect depends on the size, morphology, and volume fraction of the precipitate phases. Because the radius of Ce atoms (0.182 nm) is larger than that of Al atoms (0.143 nm), it is easy to lead to lattice distortion and increasing of the energy, when Ce atoms solute to the Al matrix in the form of a supersaturated solid solution. After solid solution treatment, many supersaturated vacancies are accumulated around Ce atoms, which results in reducing the lattice distortion energy and vacancy formation energy at a low energy level [30]. The vacancy clusters are located in the preferential nucleation positions of the primary phase T_1 due to introducing the dislocations [31]. Therefore, the nucleation of T_1 phase is promoted by the Ce atomic clusters, as shown in Figure 12.

The Ce element usually exists in the form of solid solution, and segregates at grains and phase boundaries, and rare earth compounds in Al-Cu-Li alloys. From Figure 7, it can be observed that the density of dislocations increased when 0.1% Ce was added into the AL1 alloy, and thus the contributions of dislocation strengthening increased. Additionally, the coarse dendrites were refined, and the secondary dendrites arm spacing was reduced with the increase of Ce content. The grains of T6 treated alloys are also refined, as shown in Figures 2 and 8. This is because the solid solubility of Ce element in Al alloys is very small. In the process of solidification, Ce elements easily gather at the front of the solid–liquid interface. This results in the redistribution of solute elements and increases the composition supercooling of the liquid phase, thus intensifying the branching process, refining the dendrite shape, and reducing the dendrite spacing. In addition, some compounds formed from Ce, Al, and other elements hinder the grain growth at grain boundary. Some research shows that the second Al_8Cu_4Ce particles form during solidification of Al-Cu-Ce alloys, which can promote the non-uniform nucleation [32]. The Al_8Cu_4Ce particles are broken due to the rolling process, but these broken particles still cannot be completely dissolved during the solid solution treatment, which can hinder the grain boundary migration during the subsequent recrystallization process, thus refining the recrystallized grain. The relationship between the strength increasement and grain refinement can be expressed by the following equation [33]:

$$\Delta\sigma_{H-P} = k\left(d_{Ce-containing} - d_{Ce-free}\right) \quad (1)$$

The value of k for aluminum alloys is 0.04 Mpa $\times$ m$^{1/2}$, and the Equation (1) shows that the strength of the alloy can be improved by grain refinement. From Figure 13, it can be seen that the strength of AL2 alloy is higher than that of AL1 alloy.

From Figure 6, it can be observed that the fraction of HAGBs increases from 2.4% to 8.2% with the increase of Ce content. With the increase of HAGBs, the fraction of recrystallization grains increases from 0.5% to 9.1% (Figure 5). At the same time, the average misorientation angle of the alloy increases from 2.5° to 4.0°. These results indicate that the addition of Ce element can promote dynamic recrystallization of the alloy during deformation. From Figure 9, it can be clearly observed that the fraction of recrystallized grains decreases gradually with the increase of Ce content, from 96.96% of AL2 alloy to 45.47% of AL4 alloy after T6 treatment. The reason for this is that static recrystallization of the alloy during heat treatment can be hindered due to the Ce element, which can be attributed to the formation of a large number of small Al_8Cu_4Ce phases. These Al_8Cu_4Ce phases become smaller sizes after rolling deformation with large reduction. Due to the high temperature stability of Al_8Cu_4Ce phase, it is hard to redissolve during heat treatment, which hinders recrystallization by pinning dislocation and sub-grain migration [34,35].

The more Ce content of the alloy there is, the more insoluble Al_8Cu_4Ce particles are. The Al_8Cu_4Ce phase particles can refine the microstructure and hinder recrystallization of the alloys after adding little amounts of Ce. To a certain extent, it is beneficial to improve the mechanical property of the alloys. However, with the increase of Ce content, many Cu atoms of the alloy exist in the form of Al_8Cu_4Ce particles after solid solution treatment, rather than in the solid solution matrix as solid solution Cu atoms, which leads to decreasing the mechanical properties of AL3 alloy and AL4 alloy. The addition of trace amounts of Ce can improve the binding energy of Cu atoms in the matrix. The Ce element in the matrix reduces the diffusivity of solute atoms Cu and Mg during aging and effectively inhibits the coplanar slip, which improves the mechanical property of the alloy.

4. Conclusions

The purpose of this study is to investigate the effect of Ce alloying on the microstructure and mechanical properties of Al-Cu-Li alloys. The following conclusions are drawn from the present study:

(1) The dendrites are refined, and the secondary dendrites arm spacing is reduced by the as-cast alloys with the increase of Ce content. The average recrystallized grain size of four alloys after T6 treatment are 51µm, 44µm, 43µm, and 39µm, respectively, when

Ce content is added from 0.0% to 0.3%. The Al_8Cu_4Ce particles can promote nucleation during solidification, while the particles are difficult to dissolve into the solid solution matrix during subsequent treatment, which can hinder grain boundary migration during recrystallization and refine the recrystallized grains.

(2) The recrystallization of the alloys is inhibited during T6 treatment with the increase of Ce content. This can be attributed to the existence of the Al_8Cu_4Ce phases at the grain boundary. The aging precipitation T_1 and θ' phases of the alloys are promoted by adding Ce element to the alloy.

(3) With the addition of Ce content from 0.1% to 0.3%, the ultimate tensile strength and yield strength of the alloy after T6 treatment firstly increase and then decrease. The ultimate tensile strength is a maximum of 514 MPa when Ce content is 0.1%.

Author Contributions: Conceptualization, Y.L.; methodology, J.W.; software, X.D.; validation, X.D. and Y.L.; formal analysis, D.Z.; investigation, X.D.; resources, Y.L. and J.W.; data curation, X.D. and X.L.; writing—original draft preparation, X.D.; writing—review and editing, X.D., Y.L., and J.W.; visualization, X.D.; supervision, Y.L.; project administration, Y.L., J.W.; funding acquisition, Y.L. All authors have read and agreed to the published version of the manuscript.

Funding: This research is supported by the National Science Foundation of China (No.51975404), the Postgraduate Research & Practice Innovation Program of Jiangsu Province (SJCX21_1304), the Changzhou Sci&Tech Program (CJ20220061), and the Major Project of Education Department, Jiangsu Province (No.18KJA430007).

Institutional Review Board Statement: Not applicable.

Informed Consent Statement: Informed consent was obtained from all subjects involved in the study.

Data Availability Statement: Not applicable.

Conflicts of Interest: The authors declare no conflict of interest.

References

1. Rioja, R.J.; Liu, J. The evolution of Al-Li base products for aerospace and space applications. *Metall. Mater. Trans. A* **2012**, *43*, 3325–3337. [CrossRef]
2. Williams, J.C.; Starke, E.A., Jr. Progress in structural materials for aerospace systems. *Acta Mater.* **2003**, *51*, 5775–5799. [CrossRef]
3. Araullo-Peters, V.; Gault, B.; de Geuser, F.; Deschamps, A.; Cairney, J.M. Microstructural evolution during ageing of Al–Cu–Li–x alloys. *Acta Mater.* **2014**, *66*, 199–208. [CrossRef]
4. Abd El-Aty, A.; Xu, Y.; Guo, X.; Zhang, S.-H.; Ma, Y.; Chen, D. Strengthening mechanisms, deformation behavior, and anisotropic mechanical properties of Al-Li alloys: A Review. *J. Adv. Res.* **2018**, *10*, 49–67. [CrossRef]
5. Ma, P.P.; Zhan, L.H.; Liu, C.H.; Wang, Q.; Li, H.; Liu, D.B.; Hu, Z.G. Pre-strain-dependent natural ageing and its effect on subsequent artificial ageing of an Al-Cu-Li alloy. *J. Alloys Compd.* **2019**, *790*, 8–19. [CrossRef]
6. Peng, Z.W.; Li, J.F.; Sang, F.J.; Chen, Y.L.; Zhang, X.H.; Zheng, Z.Q.; Pan, Q.L. Structures and tensile properties of Sc-containing 1445 Al-Li alloy sheet. *J. Alloys Compd.* **2018**, *747*, 471–483. [CrossRef]
7. Zhang, J.; Liu, Z.M.; Shi, D.F. Hot compression deformation behavior and microstructure of as-cast and homogenized AA2195 Al-Li alloy. *Metals* **2022**, *12*, 1580. [CrossRef]
8. Meng, L.; Zheng, X.L. Tension characteristics of notched specimens for Al-Li-Cu-Zr alloy sheets with various cerium contents. *Metall. Mater. Trans. A* **1996**, *27*, 3089–3094. [CrossRef]
9. Suresh, M.; Sharma, A.; More, A.M.; Nayan, N.; Suwas, S. Effect of Scandium addition on evolution of microstructure, texture and mechanical properties of thermo-mechanically processed Al-Li alloy AA2195. *J. Alloys Compd.* **2018**, *740*, 364–374. [CrossRef]
10. Yu, T.F.; Li, B.C.; Medjahed, A.; Hou, L.G.; Wu, R.Z.; Zhang, J.H.; Sun, J.F.; Zhang, M.L. Impeding effect of the Al_3(Er,Zr,Li) particles on planar slip and intergranular fracture mechanism of Al-3Li-1Cu-0.1Zr-X alloys. *Mater. Charact.* **2019**, *147*, 146–154. [CrossRef]
11. Liu, T.; Dong, Q.; Fu, Y.N.; Yang, J.; Zhang, J.; Sun, B.D. Effect of addition of La and Ce on solidification behavior of Al-Cu alloys. *Mater. Lett.* **2022**, *324*, 132653. [CrossRef]
12. Xiao, D.H.; Wang, J.N.; Ding, D.Y.; Yang, H.L. Effect of rare earth Ce addition on the microstructure and mechanical properties of an Al–Cu–Mg–Ag alloy. *J. Alloys Compd.* **2003**, *352*, 84–88. [CrossRef]
13. Zakharov, V.V. Special features of crystallization of scandium-alloyed aluminum alloys. *Met. Sci. Heat Treat.* **2012**, *53*, 414–419. [CrossRef]
14. Belov, N.A.; Khvan, A.V.; Alabin, A.N. Microstructure and phase composition of Al–Ce–Cu Alloys in the Al-Rich corner. *Mater. Sci. Forum* **2006**, *519–521*, 395–400. [CrossRef]

15. Belov, N.A.; Khvan, A.V. The ternary Al–Ce–Cu phase diagram in the aluminum-rich corner. *Acta Mater.* **2007**, *55*, 5473–5482. [CrossRef]
16. Bo, H.; Jin, S.; Zhang, L.G.; Chen, X.M.; Chen, H.M.; Liu, L.B.; Zheng, F.; Jin, Z.P. Thermodynamic assessment of Al–Ce–Cu system. *J. Alloys Compd.* **2009**, *484*, 286–295. [CrossRef]
17. Chaubey, A.K.; Mohapatra, S.; Jayasankar, K.; Pradhan, S.K.; Satpati, B.; Sahay, S.S.; Mishra, B.K.; Mukherjee, P.S. Effect of cerium addition on microstructure and mechanical properties of Al-Zn-Mg-Cu alloy. *Trans. Indian Inst. Met.* **2009**, *62*, 539–543. [CrossRef]
18. Yu, X.X.; Yin, D.F.; Yu, Z.M. Effects of cerium and zirconium microalloying addition on the microstructures and tensile properties of novel Al-Cu-Li alloys. *Rare Met. Mater. Eng.* **2016**, *45*, 1917–1923.
19. Ma, Y.L.; Li, J.F. Variation of aging precipitates and mechanical strength of Al-Cu-Li alloys caused by small addition of rare earth elements. *J. Mater. Eng. Perform.* **2017**, *26*, 4329–4339. [CrossRef]
20. Fang, H.C.; Chen, K.H.; Zhang, Z.; Zhu, C.J. Effect of Yb additions on microstructures and properties of 7A60 aluminum alloy. *Trans. Nonferrous Met. Soc. China* **2008**, *18*, 28–32. [CrossRef]
21. Fang, H.C.; Chao, H.; Chen, K.H. Effect of recrystallization on intergranular fracture and corrosion of Al–Zn–Mg–Cu–Zr alloy. *J. Alloys Compd.* **2015**, *622*, 166–173. [CrossRef]
22. Qin, C.; Gou, G.Q.; Che, X.L.; Chen, H.; Chen, J.; Li, P.; Gao, W. Effect of composition on tensile properties and fracture toughness of Al–Zn–Mg alloy (A7N01S-T5) used in high speed trains. *Mater. Des.* **2016**, *91*, 278–285. [CrossRef]
23. Cong, F.G.; Zhao, G.; Jiang, F.; Tian, N.; Li, R.F. Effect of homogenization treatment on microstructure and mechanical properties of DC cast 7X50 aluminum alloy. *Trans. Nonferrous Met. Soc. China* **2015**, *25*, 1027–1034. [CrossRef]
24. Pasang, T.; Symonds, N.; Moutsos, S.; Wanhill, R.J.H.; Lynch, S.P. Low-energy intergranular fracture in Al–Li alloys. *Eng. Fail. Anal.* **2012**, *22*, 166–178. [CrossRef]
25. Ma, J.; Yan, D.S.; Rong, L.J.; Li, Y.Y. Effect of Sc addition on microstructure and mechanical properties of 1460 alloy. *Prog. Nat. Sci. Mater. Int.* **2014**, *24*, 13–18. [CrossRef]
26. Yu, X.X.; Dai, H.; Li, Z.T.; Sun, J.; Zhao, J.F.; Li, C.Q.; Liu, W.W. Improved recrystallization resistance of Al–Cu–Li–Zr alloy through Ce addition. *Metals* **2018**, *8*, 1035. [CrossRef]
27. Zuo, J.R.; Hou, L.G.; Shi, J.T.; Cui, H.; Zhuang, L.Z.; Zhang, J.S. The mechanism of grain refinement and plasticity enhancement by an improved thermomechanical treatment of 7055 Al alloy. *Mater. Sci. Eng. A* **2017**, *702*, 42–52. [CrossRef]
28. da Costa Teixeira, J.; Cram, D.G.; Bourgeois, L.; Bastow, T.J.; Hill, A.J.; Hutchinson, C.R. On the strengthening response of aluminum alloys containing shear-resistant plate-shaped precipitates. *Acta Mater.* **2008**, *56*, 6109–6122. [CrossRef]
29. Deschamps, A.; Decreus, B.; De Geuser, F.; Dorin, T.; Weyland, M. The influence of precipitation on plastic deformation of Al–Cu–Li alloys. *Acta Mater.* **2013**, *61*, 4010–4021. [CrossRef]
30. Wang, W.T.; Zhang, X.M.; Gao, Z.G.; Jia, Y.Z.; Ye, L.Y.; Zheng, D.W.; Liu, L. Influences of Ce addition on the microstructures and mechanical properties of 2519A aluminum alloy plate. *J. Alloys Compd.* **2010**, *491*, 366–371. [CrossRef]
31. Kumar, K.S.; Brown, S.A.; Pickens, J.R. Microstructural evolution during aging of an Al-Cu-Li-Ag-Mg-Zr alloy. *Acta Mater.* **1996**, *44*, 1899–1915. [CrossRef]
32. Yu, X.X.; Yin, D.F.; Yu, Z.M.; Zhang, Y.R.; Li, S.F. Effects of cerium addition on solidification behaviour and intermetallic structure of novel Al-Cu-Li alloys. *Rare Met. Mater. Eng.* **2016**, *45*, 1423–1429.
33. Hansen, N. Hall–Petch relation and boundary strengthening. *Scr. Mater.* **2004**, *51*, 801–806. [CrossRef]
34. Yu, X.X.; Yin, D.F.; Yu, Z.M.; Zhang, Y.R.; Li, S.F. Microstructure evolution of novel Al-Cu-Li-Ce alloys during homogenization. *Rare Met. Mater. Eng.* **2016**, *45*, 1687–1694.
35. Balducci, E.; Ceschini, L.; Messieri, S.; Wenner, S.; Holmestad, R. Thermal stability of the lightweight 2099 Al-Cu-Li alloy: Tensile tests and microstructural investigations after overaging. *Mater. Des.* **2017**, *119*, 54–64. [CrossRef]

Disclaimer/Publisher's Note: The statements, opinions and data contained in all publications are solely those of the individual author(s) and contributor(s) and not of MDPI and/or the editor(s). MDPI and/or the editor(s) disclaim responsibility for any injury to people or property resulting from any ideas, methods, instructions or products referred to in the content.

Article

Shear Transformation Zone and Its Correlation with Fracture Characteristics for Fe-Based Amorphous Ribbons in Different Structural States

Weiwei Dong *, Minshuai Dong, Danbo Qian, Jiankang Zhang and Shigen Zhu

College of Mechanical Engineering, Donghua University, Shanghai 201620, China
* Correspondence: wwddhush@dhu.edu.cn

Abstract: Fe-based amorphous alloys often exhibit severe brittleness induced by annealing treatment, which increases the difficulties in handling and application in the industry. In this work, the shear transformation zone and its correlation with fracture characteristics for FeSiB amorphous alloy ribbons in different structural states were investigated. The results show that the bending strain decreases sharply with the annealing temperature increase, accompanied by decreased shear band density and the induced plastic deformation zone. Furthermore, the microscopic fracture surface features transform from a micron-scale dimple pattern to nano-scale dimples and periodic corrugations. According to nano-indentation results, the strain rate sensitivity and shear transformation zone volume change significantly upon annealing treatment, which is responsible for the deterioration of bending ductility and the transition of microscopic fracture surface features.

Keywords: shear band; shear transformation zone; fracture surface; dimple pattern; periodic corrugations

1. Introduction

FeSiB amorphous alloy ribbons are a widely used soft magnetic material, which is characterized by its simple preparation process, low production cost, and low loss rate. In addition, it exhibits excellent properties such as high magnetic permeability, low coercive force, and low iron core loss, making it widely applied in the iron cores of commercial high-efficiency distribution transformers, intermediate frequency (400–10,000 Hz) transformers, filters, inductors, and reactors. Therefore, optimizing the soft magnetic properties of FeSiB amorphous alloy ribbon is a hot research topic in this field. Among these, the annealing treatment of FeSiB amorphous alloy ribbon below the crystallization temperature is an effective method, which can further improve the soft magnetic properties of Fe-based amorphous alloys through structure relaxation. However, FeSiB amorphous alloy ribbon often suffers from severe annealing embrittlement, which affects its comprehensive properties, increases its processing and application difficulties in the industry, and has attracted high attention from the academic and industrial communities [1,2].

The plastic deformation mechanism of amorphous alloys is completely different from that of crystalline materials because the atomic arrangement of amorphous alloys exhibits short-range order and long-range disorder, and there are no crystal defects such as dislocations and grain boundaries. Therefore, the plastic deformation theory of traditional crystalline materials cannot be applied to amorphous alloys. Currently, the main theories of amorphous alloy deformation are the free volume theory and the shear transformation zone (STZ) theory. The free volume model is a phenomenological physical concept, which states that plastic deformation of amorphous alloys can be achieved under local atomic transition conditions. After annealing treatment, the free volume of amorphous alloys is eliminated due to the random jump motion of atoms, leading to the structural relaxation of amorphous alloys. The reduction in free volume after annealing decreases the plastic deformation ability of amorphous alloys [3,4]. However, the microstructural origin of

Citation: Dong, W.; Dong, M.; Qian, D.; Zhang, J.; Zhu, S. Shear Transformation Zone and Its Correlation with Fracture Characteristics for Fe-Based Amorphous Ribbons in Different Structural States. *Metals* **2023**, *13*, 757. https://doi.org/10.3390/met13040757

Academic Editors: Sundeep Mukherjee and Giovanni Meneghetti

Received: 26 February 2023
Revised: 1 April 2023
Accepted: 10 April 2023
Published: 13 April 2023

Copyright: © 2023 by the authors. Licensee MDPI, Basel, Switzerland. This article is an open access article distributed under the terms and conditions of the Creative Commons Attribution (CC BY) license (https://creativecommons.org/licenses/by/4.0/).

embrittlement caused by structural relaxation is not clear yet. Currently, a large amount of research has shown that the plastic deformation mechanism of amorphous alloys is closely related to shear bands. Shear bands are a typical feature of ductile banded amorphous alloys, which are generated when ductile bands start to bend, and their formation is related to the evolution of local structural order. When amorphous alloy ribbons have a large number of shear bands, it indicates that they have good bending ductility. In the plastic deformation process of amorphous alloys, the activation of STZs is an important step. The STZ theory states that the plastic deformation of amorphous alloys is not realized by the transition of single atoms but by clusters of atoms composed of numerous atoms as flow units, under certain conditions. STZ is a micro deformation area caused by local structural changes, and its size is usually between several nanometers to tens of nanometers. The activation of STZs in amorphous alloys is closely related to factors such as material composition, temperature, strain rate, and stress level [5–7]. Therefore, understanding the plastic deformation mechanism of amorphous alloys and the formation of shear bands has important academic value and practical significance for the development of high-performance amorphous alloy materials.

STZ has been widely employed to study the low-temperature deformation of amorphous alloys. However, the correlation between the measured STZ size and amorphous alloys ductility was also controversial. Particularly, Pan et al. [8,9] evaluated the STZs size in amorphous alloy by using the rate-change method. It was found that upon annealing, the measured STZ size dramatically decreases. The study demonstrates that amorphous alloy with larger STZ size enjoys better plasticity. Whereas someone held the opposite opinion that amorphous alloy with smaller STZ size possesses better plasticity. Ma et al. [10] employed statistical analysis of the maximum shear stress, which is based on the distribution of the first pop-in events in nanoindentation, to estimate the STZ volume and atoms it contains in the amorphous alloy. The result indicates that the STZ size decreases from 83 atoms in the ribbon to 36 atoms in the film. Upon nanoindentation creep test, STZ sizes with 44 atoms and 18 atoms were calculated for the ribbon and film, respectively. Choi et al. [11] employed statistical analysis of the data to estimate the STZ size. The results of the analysis indicate an STZ size of ~25 atoms in amorphous alloy, which increases to ~34 atoms upon annealing. The activation volume of a single STZ was estimated from either a statistical analysis or the rate-change method. The STZ size obtained by the rate-change method shows a better correlation with amorphous alloy plasticity [12,13].

In order to reveal the mechanism of brittleness for FeSiB amorphous alloy ribbons, this study was carried out. To investigate the relationship between shear transformation zone (STZ) size and bending ductility of FeSiB amorphous alloy ribbons, the FeSiB amorphous alloy ribbons were annealed at various temperatures. FeSiB amorphous alloy ribbons with different structural states were selected as the research objects. In this study, the STZ size was measured by the rate change method, and the bending toughness, folding morphology, and resulting fracture surfaces of FeSiB amorphous ribbons under different free volume states were further studied. The formation mechanism of shear bands and fracture surface characteristics were systematically discussed. The results showed that reducing the volume of STZs can effectively reduce the brittleness of FeSiB amorphous alloy ribbons, providing new ideas for improving the bending ductility of FeSiB amorphous alloys by controlling their structural states. Meanwhile, this study has important theoretical significance for a deeper understanding of the fracture behavior of amorphous alloys and its relationship with structural factors and can provide certain technical guidance for the development of high-performance amorphous alloys.

2. Materials and Methods

All materials used in this experiment are of high purity. The mass of each element was calculated based on the nominal composition of $Fe_{80}Si_9B_{11}$, and the required raw materials were weighed using a high-precision electronic balance. Prior to melting the high-purity raw materials, titanium ingots were melted to absorb oxygen and impurities in the high-vacuum arc furnace. The furnace vacuum degree was reduced to below 5×10^{-3} Pa and high-purity argon gas was introduced for protection during the melting process. Under the heating conditions of an induction coil, a mixture of pure Fe (99.9 mass%), Si (99.9 mass%), and FeB (17.9 mass%B) was melted to prepare an Fe80Si9B11 alloy ingot. The ingot was repeatedly melted four times and finally cooled for 30 min with furnace cooling to obtain a homogenous master alloy ingot. Before preparing the FeSiB amorphous alloy ribbons, the surface impurities of the master alloy ingot were thoroughly removed and subsequently shattered into small alloy ingots. The single-roller melt-spinning technique was used to prepare the amorphous alloy strip, which is easy to operate and has a fast cooling rate, making it possible to produce suitable amorphous alloy strips. The preparation process of the FeSiB amorphous alloy ribbons was carried out completely in a high-purity argon gas atmosphere, with an argon pressure differential of 0.02 MPa to ensure a stable atmosphere. The molten alloy was ejected from the narrow slit at the bottom of a quartz tube onto a rapidly rotating copper wheel, with a surface speed of 40 m/s, allowing for control of the thickness and width of the amorphous alloy ribbons. The ribbons were 25 ± 2 µm in thickness and 8 mm in width. The structure and morphology of the ribbons were examined using X-ray diffraction (XRD) with Co-Kα (λ = 1.7902 Å) under a step of 0.01° and a counting time of 4 s per step at 40 kV and 20 mA. The as-cast and annealed at 250 °C and 380 °C amorphous ribbons were labeled as S25, S250, and S380, respectively. The S250 and S380 ribbons were annealed at 250 and 380 °C for 90 min in a tubular vacuum furnace (<0.1 Pa) with furnace cooling, the heating rate was 10 °C/min. The bending ductility of ribbons was tested with a self-designed two-point bending device, which was composed of three boards. The schematic diagram of the two-point bending test is shown in Figure 1. The top and side boards were fixed, and the lower one could move from up and down. The ribbons were fixed on the upper and lower boards, and then the lower board was controlled to move up at the speed of 0.1 mm/s. The bending strain was used as ductility index λ_f to value the bending ductility of the ribbons [1], as Equation (1):

$$\lambda_f = \frac{t}{L_f - t} \quad (1)$$

where t is the ribbon's thickness, and L_f is the board spacing when the ribbons have just broken. The ribbons which could be folded at 180° without breaking down show good ductility.

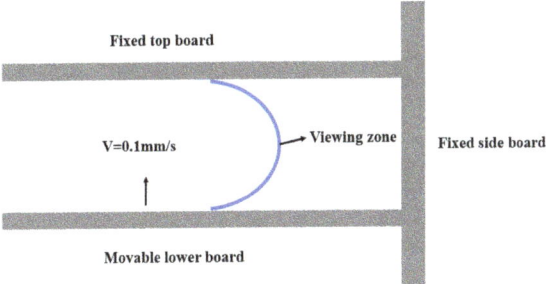

Figure 1. The schematic diagram of two-point bending test.

The macro and micro images of the bent crease and resultant fracture surface of these three samples were investigated by scanning electron microscopy (Field Emission Scanning Electron Microscope, FESEM, JSM-7500F, JEOL, Japan) and transmission electron microscope (TEM, JEM-2100, JEOL, Japan). The FeSiB amorphous alloy ribbons were subjected to nano-indentation testing (Agilent G200, Keysight, Santa Rosa, CA, USA) using a Berkovich diamond indenter with an effective tip radius of 25 nm. The experimental samples were mechanically polished to ensure that the surface of the FeSiB amorphous alloy ribbons was perpendicular to the loading direction of the indenter and tested in a constant temperature and humidity as well as a quiet environment. The load was increased at rates of 0.01, 0.05, 0.1, and 0.2 mN/s. The nano-indentation test was conducted 10 times at various loading rates for each sample.

3. Results

Results and Discussion

As shown in Figure 2, the XRD patterns of as-cast and annealed FeSiB alloy samples S25, S250, and S380 exhibit broad peaks without any sharp diffraction peak related to the crystalline phase. This suggests that all the samples keep an amorphous structure.

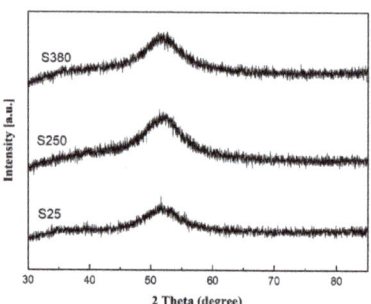

Figure 2. XRD patterns for amorphous alloy ribbons in different structural states: S25, S250, and S380.

In order to further confirm that the as-cast and annealed FeSiB amorphous alloy ribbons are amorphous, they were tested by HRTEM. Figure 3 shows the HRTEM images and corresponding selected area electron diffraction (SAED) patterns of S25 and S380. For the as-cast (Figure 3a) and annealed sample (Figure 3b), no precipitated nanometer-sized phase is detected, only the homogeneous mazy contrast and a diffuse halo are presented, This is a typical structural characteristic of amorphous alloys, which is consistent with XRD results, demonstrating that these three samples annealed at different temperatures still remain a monolithic amorphous structure.

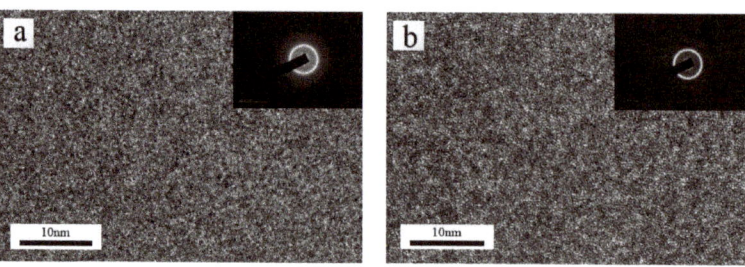

Figure 3. High-resolution TEM image inset with corresponding SAED pattern for (**a**) S25 and (**b**) S380.

It is well known that amorphous alloy is usually located in a metastable state, a structure relaxation would process this in a wide temperature range from room temperature to crystallization temperature. The exothermic event accompanied by structure relaxation shows that the amount of free volume in amorphous alloys is decreased. The exothermic enthalpy value is related to the free volume content,

$$\Delta H_{f_v} = \beta \cdot \Delta V_f \qquad (2)$$

where ΔH is exothermic enthalpy, which is positive to the annealing temperature, ΔV_f is the change of free volume, β is a constant parameter [14]. In this study, for these three samples, the exothermic enthalpy values of samples S250 and S380 increase with the increase in the annealing temperature. It suggests that the as-cast sample S25 contains the largest ΔV_f, and sample S380 has the least free volume left.

The two-point bending test and SEM were used to characterize the bending ductility and bent crease morphology of FeSiB alloy ribbons. The samples S25 and S250 show better bending ductility than S380. The bending strain of S25 and S250 both are 1 but S25 could be bent to a 180-degree angle about 10 times, and the bend times of S250 decrease to about 3 times. Meanwhile, the S380 exhibits no bending ductility and will be broken down in the initial stage of the bending test. Since the shear bands, owing to the plastic strain of amorphous alloy ribbons, could reflect the plastic deformation ability, a significant difference in shear band behavior can be noticed in Figure 4. S25 exhibits the most shear bands, followed by S250, and S380 shows the least. For S25 and S250, the shear bands are parallel to the bent crease and propagate in a normal direction to the bent crease. However, there are two obvious fracture edges and no shear bands for S380. The center area, where the distance is plus or minus 5μm from the bent crease, is selected to show the number of shear bands. For S25, the average number of shear bands is around 25, while it decreased to 5 for S250, then to 0 for S380. Moreover, several distinct cracks are produced in the bent crease for S250, suggesting decreased bending ductility, which is in agreement with the bending strain results.

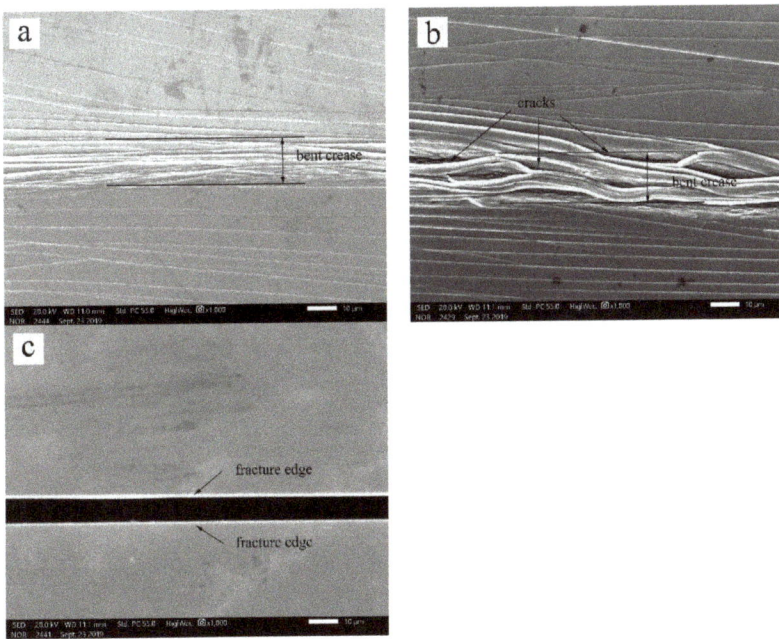

Figure 4. Shear band morphologies for amorphous alloy ribbons annealed at different temperatures: (**a**) S25; (**b**) S250; and (**c**) S380.

The fracture surface for the samples in different structure states was investigated with SEM, as shown in Figures 5–7. The fracture surface of S25 shows a periodic morphology consisting of a micro-scaled dimple pattern zone and the largest smooth zone, revealing ductile features (Figure 5a). The share of dimple pattern zone is up to 70%, the average dimple diameter is about 500 nm, and the smooth zone displacement is about 3.22 μm (Figure 5b,c). For S250, the fracture surface consists of a micro-scaled dimple pattern zone, smooth zone, and mist zone (Figure 6a), suggesting local ductile fracture. The share of dimple pattern zone decreases to 20%, the average dimple diameter is about 200 nm, and smooth zone displacement is about 1.52 μm (Figure 6b,d). At the microscale, the fracture surface of S380 consists only of the smooth zone as shown in Figure 7a, suggesting a completely brittle fracture. From the high magnification image Figure 7b,c, both smooth zone and nanoscale configuration patterns appear on the fracture surface. The average dimple sizes further decrease to 100 nm and the smooth zone displacement is only about 500 nm (Figure 7b,d). The fractograph evolution from a micro-scaled dimple-like structure to a nano-scaled dimple-like structure and then to the periodic corrugation pattern along the crack propagation direction, indicates a ductile to brittle transition during the dynamic fracture process of S250. The fractograph of S380 consists only of the smooth zone. This means that the intrinsic plasticity of S250 and S380 have little resistance to crack propagation, and cracks will propagate rapidly throughout the whole sample until fracture once initiated. Together with the previous bending test, it can be concluded that the micro-scaled dimple pattern zone corresponds to the ductile feature, and those nano-scale dimples and periodic corrugations are typical fracture features of the quasi-brittle fracture of metallic glasses [15–18].

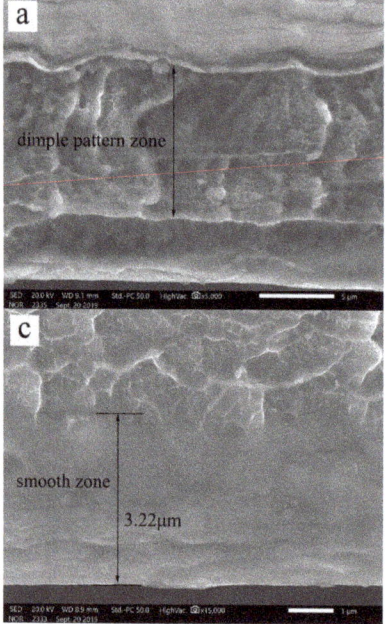
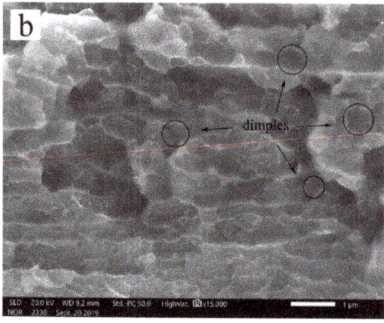

Figure 5. Fracture surfaces for S25: micro-scaled dimple pattern zone (**a**); enlarged images for dimple (**b**); smooth zone (**c**).

As reported, the size of STZ had been suggested as a key element in such plastic deformation [7,19,20]. The STZ volume, Ω, obtained by the rate-change method can be expressed as:

$$\Omega = kT/C'mH \tag{3}$$

where k is the Boltzmann constant, T is the temperature, H is the hardness, m is the corresponding strain rate sensitivity of hardness.

$$C' = \frac{2R\xi}{\sqrt{3}} \frac{G_0 \gamma_C^2}{\tau_C} \left(1 - \frac{\tau_{CT}}{\tau_C}\right)^{1/2} \tag{4}$$

Here, $R \approx 1/4$ and $\xi \approx 3$ are constants, τ_c is the threshold shear resistance at temperature T and the $\tau_c/G_0 \approx 0.036$. The average elastic limit g $\gamma_c \approx 0.027$ and the value of τ_{ct}/τ_c can be determined by $\tau_{ct}/G = \gamma_{c0} - \gamma_{c1}(T/T_g)^{2/3}$, where $\gamma_{c0} = 0.036 \pm 0.002$, $\gamma_{c1} = 0.016 \pm 0.002$ [21]. In order to obtain the STZ volume, calculating m is essential.

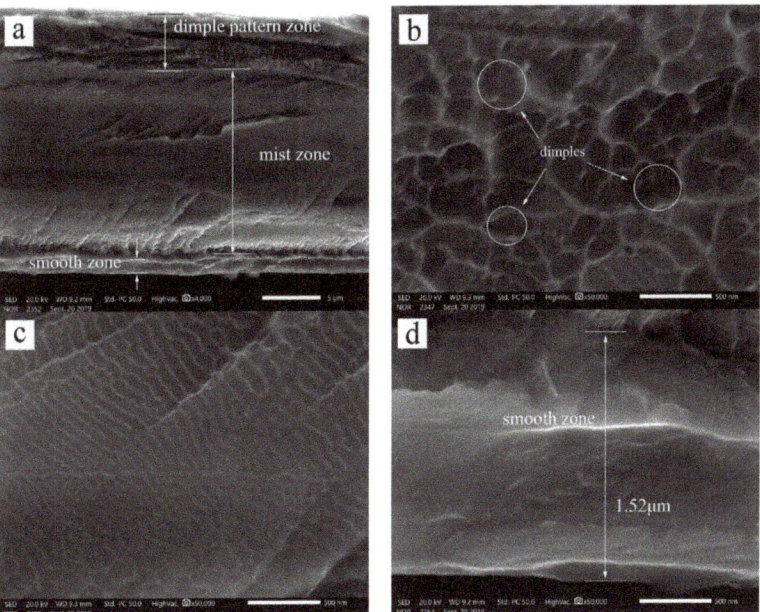

Figure 6. Fracture surfaces for S250: overview of fracture surface (**a**); enlarged images for dimple (**b**); mist zone (**c**); smooth zone (**d**).

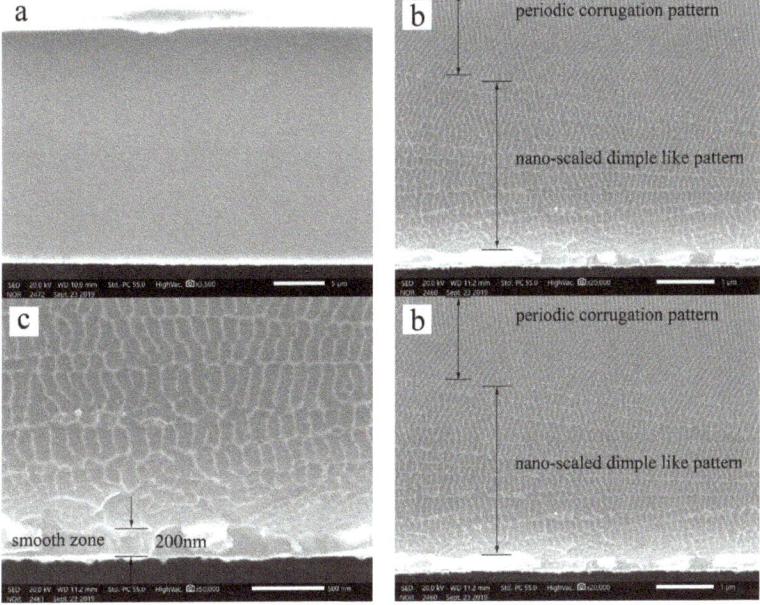

Figure 7. Fracture surfaces for S380: overview of fracture surface (**a**); enlarged images for nano-scaled dimple-like pattern (**b**); smooth zone (**c**); mist zone (**d**).

To investigate the changes in the bending ductility of FeSiB amorphous alloy ribbons after annealing treatment, nanoindentation experiments were conducted. The typical force-depth curves and hardness of S25, S250, and S380 by nano-indentation with a loading rate of 0.2 mN/s are shown in Figure 8. The maximum indentation depths for samples S380 and S25 were 52 nm and 75 nm, respectively. It was observed that the maximum indentation depth for S380 was significantly lower than that for S25, and the deformation rate of FeSiB amorphous alloy ribbons decreased continuously with increasing annealing temperature. In addition, the average hardness of S25 was 5.393 GPa, and the average hardness of FeSiB amorphous alloy ribbons increased continuously with increasing annealing temperature. Specifically, the average hardness of S250 and S380 increased to 6.992 GPa and 8.472 GPa, respectively. To calculate the value of m, the force-depth curves of S25, S250, and S380 by nano-indentation under loading rates of 0.01, 0.05, 0.1, and 0.2 mN/s are tested. According to Pan [9], the strain rate sensitivity m is proportional to the ratio of double log plot of strain rate $\dot{\varepsilon}$ to hardness H. The strain rate ε = P/2P, where P is the loading rate and P is the peak load force. The m values of as-cast, 250 °C annealed, 380 °C annealed samples are 0.340, 0.228, and 0.116 (as shown in Figure 9), respectively. In order to calculate the numbers of STZ, the atomic radius, based on the dense-packing hard-sphere model, can be computed as:

$$R = \left(\sum_i^n A_i r_i^3\right)^{\frac{1}{3}} \quad (5)$$

where A_i and r_i are the atomic fraction and atomic radius of each element, respectively [13]. The STZ volume, Ω, and the number of atoms, N, in the STZ are calculated and summarized in Table 1. It can be seen that U and N increase sharply with the annealing temperature. The Ω increases from 0.54 nm^3 for S25 to 0.63 nm^3 for S250 and 1.01 nm^3 for S380, respectively. The N, in the STZ of the S25, S250, and S380 is estimated to be 66, 76, and 123, respectively. It can be found that the variation tendency of N is the same as STZ volume. It has been accepted that densely packed atomic clusters and loosely packed liquid-like regions in amorphous alloys participate in plastic flow under deformation, which is thought to be the root of STZ events [12,22]. In general, the free volume will annihilate, liquid-like regions will be fewer and smaller in structural relaxation during the annealing process, and a more uniform and stable microstructure will be obtained. The liquid-like zones, in which the atom numbers could be changed under the stress, are able to participate in plastic flow, resulting in rising STZ events. Once an STZ forms, another one may generate in the neighborhood by the local strain field. As the stress spreads, more and more STZ would be activated until a shear band is created. In addition, the new STZs can be activated due to the larger local strain field caused by the adjacent shear band.

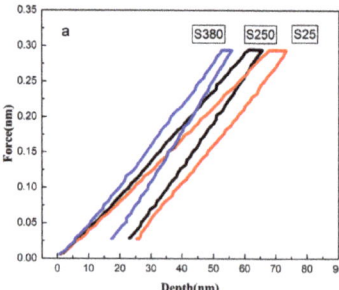
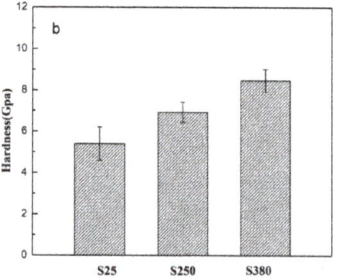

Figure 8. Force-depth curves (**a**) and hardness (**b**) of S25, S250, and S380 by nanoindentation with a loading rate of 0.2 mN/s.

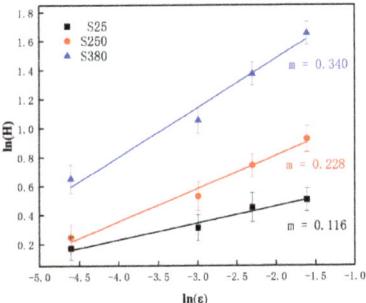

Figure 9. Determination of the strain rate of hardness(m) for S25, S250, and S380.

Table 1. The detailed measurement data on STZ for various samples.

Sample	Average Hardness H (Gpa)	Strain Rate Sensitivity, m	STZ Volume Ω (nm^3)	STZ Size N (Atoms)
S25	5.393	0.340	0.54	66
S250	6.922	0.228	0.63	76
S380	8.472	0.116	1.01	123

As reported, the larger Ω is, the less activated STZ is required to form a shear band [23]. In this study, S25 and S250 exhibit larger Ω and N, so more parallel and uniformly distributed shear bands in the bending crease exist in these two samples, showing better bending ductility.

During the bending process, the hydrostatic stress is max located on the tensile side of the ribbons, and the smooth region is formed by a sliding shear for crack nucleation occurring in shear bands.

At the prime stage of crack propagation, the crack tip slowly advances into the fracture process zone, and a large stress field is built. Once the crack length and the internal stress in the vicinity of the crack tip reach the critical value, the internal stress drives the crack to propagate in the adjacent area [24,25]. The amorphous matrix in front of the crack tip, containing numerous STZ and free volume, has enough time to rearrange. With the crack branching, the free volume supplies enough space for STZ rearrangement, and the dimple structure forms in the front of the crack tip. As a result, the periodical fatigue fracture characteristic forms on the fracture surface.

Thus, it can be noticed that the dimple patterns follow the smooth zone, but the radial dimple pattern of 3.3 μm for S25 decreases to 1.3 μm for S250 and then to 300 nm for S380. This may be attributed to the STZ volume and its correlation with the neighboring liquid-like areas in the plastic deformation zone ahead of the crack tips. Upon the long-range elastic field inducing the fluctuating internal stress and strain rate, larger STZ, more free volume, and liquid-like areas take part in the plastic flow and give rise to larger and deeper cavities. The largest critical shear displacement, dimple pattern size, and radial dimple pattern imply that S25 suffers lower crack propagation speed after crack initiation. The reduced critical shear displacement, dimple pattern size, and radial dimple pattern suggest that the crack propagation speed is not fixed. The crack propagation speed decreases along the crack propagation direction for S250. For the lowest STZ size and free volume in S380, the viscoelastic medium acts as an elastic-like deformation, and the internal stress is built up in the compressed viscoelastic medium. The STZ in the fracture process zone before the crack tip has no time to rearrange and the cavities are compressed. The nano-scaled dimple pattern appears instead of the micro-scaled dimple pattern. When the fracture zone is less than the wavelength of the meniscus instability, the dimple-like structure is suppressed and the periodic corrugation pattern appears.

4. Conclusions

In this work, the shear transformation zone and its correlation with fracture characteristics for Fe-based amorphous ribbons in different structural states were studied. Samples S25 and S250 show better bending ductility and the bending strains are both 1, while S380 exhibits no bending ductility. The shear band density decreases with the decrease in free volume. There are the most shear bands on the bent crease of S25. The fractograph evolution from a micro-scaled dimple-like structure to a nano-scaled dimple-like structure and then to the periodic corrugation pattern along the crack propagation direction, indicates a ductile to brittle transition during the dynamic fracture process for ductile samples. The fracture surface of S25 shows a micro-scaled dimple pattern zone and the largest smooth zone in these three samples, and the size of the dimple pattern zone is up to 70%. The corresponding part of S380 only consists of a smooth zone, suggesting a completely brittle fracture. The better bending ductility for S25 is attributed to the smaller STZ size. The nano-intention result indicates that the strain rate sensitivity of hardness, STZ volume, and the number of atoms in STZ increase after annealing. With the STZ volume increase, the fractograph evolution to a nano-scaled dimple-like structure and then to the periodic corrugation pattern along the crack propagation direction indicates that the amorphous alloy has become embrittlement.

Author Contributions: Conceptualization, W.D.; methodology, D.Q.; software, D.Q. and J.Z.; validation, W.D.; formal analysis, M.D. and J.Z.; investigation, W.D. and D.Q.; resources, S.Z.; data curation, W.D., M.D., J.Z. and S.Z.; writing—original draft preparation, W.D, M.D., D.Q., J.Z. and S.Z.; writing—review and editing, W.D. and M.D.; visualization, W.D.; supervision, W.D. and S.Z.; project administration, W.D. and S.Z.; funding acquisition, S.Z. All authors have read and agreed to the published version of the manuscript.

Funding: This work was supported by the Chinese Universities Scientific Fund (22D110323).

Institutional Review Board Statement: Not applicable.

Informed Consent Statement: Not applicable.

Data Availability Statement: Not applicable.

Conflicts of Interest: The authors declare no conflict of interest.

References

1. Minnert, C.; Kuhnt, M.; Bruns, S. Study on the embrittlement of flash annealed $Fe_{85.2}B_{9.5}P_4Cu_{0.8}Si_{0.5}$ metallic glass ribbons. *Mater. Des.* **2018**, *156*, 252–261. [CrossRef]
2. Liang, X.; He, A.; Wang, A.; Pang, J.; Wang, C.; Chang, C. Fe content dependence of magnetic properties and bending ductility of FeSiBPC amorphous alloy ribbons. *J. Alloys Compd.* **2017**, *694*, 1260–1264. [CrossRef]
3. Dong, W.; Han, B.; Hui, J.; Yan, M. Bending behavior and fracture surface characters for FeSiB amorphous ribbons in different free volume state. *Appl. Phys. A* **2020**, *126*, 670. [CrossRef]
4. Argon, A.S. Plastic deformation in metallic glasses. *Acta Metall.* **1979**, *27*, 47–58. [CrossRef]
5. Greer, A.L.; Cheng, Y.Q.; Ma, E. Shear bands in metallic glasses. *Mater. Sci. Eng.* **2013**, *R74*, 71–132. [CrossRef]
6. Narayan, R.L.; Raut, D.; Ramamurty, U. A quantitative connection between shear band mediated plasticity and fracture initiation toughness of metallic glasses. *Acta Mater.* **2018**, *150*, 69–77. [CrossRef]
7. Jiang, F.; Jiang, M.Q.; Wang, H.F.; Zhao, Y.L.; He, L.; Sun, J. Shear transformation zone volume determining ductile–brittle transition of bulk metallic glasses. *Acta Mater.* **2011**, *59*, 2057–2068. [CrossRef]
8. Pan, D.; Yokoyama, Y.; Fujita, T.; Liu, Y.H.; Kohara, S.; Inoue, A.; Chen, M.W. Correlation between structural relaxation and shear transformation zone volume of a bulk metallic glass. *Appl. Phys. Lett.* **2009**, *95*, 141909. [CrossRef]
9. Pan, D.; Inoue, A.; Sakurai, T.; Chen, M.W. Experimental characterization of shear transformation zones for plastic flow of bulk metallic glasses. *Proc. Natl. Acad. Sci. USA* **2008**, *105*, 14769–14772. [CrossRef]
10. Ma, Y.; Ye, J.H.; Peng, G.J.; Wen, D.H.; Zhang, T.H. Nanoindentation study of size effect on shear transformation zone size in a Ni–Nb metallic glass. *Mater. Sci. Eng.* **2015**, *627*, 153–160. [CrossRef]
11. Choi, I.-C.; Zhao, Y.; Kim, Y.-J.; Yoo, B.-G.; Suh, J.-Y.; Ramamurty, U.; Jang, J.-I. Indentation size effect and shear transformation zone size in a bulk metallic glass in two different structural states. *Acta Mater.* **2012**, *60*, 6862–6868. [CrossRef]
12. Chen, Z.Q.; Huang, L.; Huang, P.; Xu, K.W.; Wang, F.; Lu, T.J. Clarification on shear transformation zone size and its correlation with plasticity for Zr-based bulk metallic glass in different structural states. *Mater. Sci. Eng. A* **2016**, *677*, 349–355. [CrossRef]

13. Ma, Y.; Peng, G.J.; Debela, T.T.; Zhang, T.H. Nanoindentation study on the characteristic of shear transformation zone volume in metallic glassy films. *Scr. Mater.* **2015**, *108*, 52–55. [CrossRef]
14. Slipenyuk, A.; Eckert, J. Correlation between enthalpy change and free volume reduction during structural relaxation of $Zr_{55}Cu_{30}Al_{10}Ni_5$ metallic glass. *J. Scr. Mater.* **2004**, *50*, 39–44. [CrossRef]
15. Raghavan, R.; Murali, P.; Ramamurty, U. On factors influencing the ductile-to-brittle transition in a bulk metallic glass. *Acta Mater.* **2009**, *57*, 3332. [CrossRef]
16. Xi, X.K.; Zhao, D.Q.; Pan, M.X.; Wang, W.H.; Wu, Y.; Lewandowski, J.J. Fracture of Brittle Metallic Glasses: Brittleness or Plasticity. *Phys. Rev. Lett.* **2005**, *94*, 125510. [CrossRef]
17. Jiang, M.Q.; Ling, Z.; Meng, J.X.; Dai, L.H. Energy dissipation in fracture of bulk metallic glasses via inherent competition between local softening and quasi-cleavage. *Philos. Mag.* **2008**, *88*, 407. [CrossRef]
18. Wang, G.; Zhao, D.Q.; Bai, H.Y.; Pan, M.X.; Xia, A.L.; Han, B.S.; Xi, X.K.; Wu, Y.; Wang, W.H. Nanoscale Periodic Morphologies on the Fracture Surface of Brittle Metallic Glasses. *Phys. Rev. Lett.* **2007**, *98*, 235501. [CrossRef]
19. Wu, F.F.; Zheng, W.; Deng, J.W.; Qu, D.D.; Shen, J. Super-high compressive plastic deformation behaviors of Zr-based metallic glass at room temperature. *Mater. Sci. Eng. A* **2012**, *541*, 199–203. [CrossRef]
20. Pan, D.; Guo, H.; Zhang, W.; Inoue, A.; Chen, M.W. Temperature-induced anomalous brittle-to-ductile transition of bulk metallic glasses. *Appl. Phys. Lett.* **2011**, *99*, 241907. [CrossRef]
21. Johnson, W.L.; Samwer, K. A universal criterion for plastic yielding of metallic glasses with a (T/Tg) 2/3 temperature dependence. *Phys. Rev. Lett.* **2005**, *95*, 195501. [CrossRef] [PubMed]
22. Liu, Y.H.; Wang, D.; Nakajima, K.; Zhang, W.; Hirata, A.; Nishi, T.; Inoue, A.; Chen, M.W. Characterization of nanoscale mechanical heterogeneity in a metallic glass by dynamic force microscopy. *Phys. Rev. Lett.* **2011**, *106*, 125504. [CrossRef] [PubMed]
23. Guo, W.; Yamada, R.; Saida, J. Rejuvenation and plasticization of metallic glass by deep cryogenic cycling treatment. *Intermetallics* **2018**, *93*, 141–147. [CrossRef]
24. Inoue, A.; Zhang, T.; Masumoto, T. Glass-forming ability of alloys. *J. Non-Cryst. Solids* **1993**, *156*, 598. [CrossRef]
25. Wu, F.F.; Zhang, Z.F.; Mao, S.X. Size-dependent shear fracture and global tensile plasticity of metallic glasses. *Acta Mater.* **2009**, *57*, 257. [CrossRef]

Disclaimer/Publisher's Note: The statements, opinions and data contained in all publications are solely those of the individual author(s) and contributor(s) and not of MDPI and/or the editor(s). MDPI and/or the editor(s) disclaim responsibility for any injury to people or property resulting from any ideas, methods, instructions or products referred to in the content.

Article

Study on Nitrogen-Doped Biomass Carbon-Based Composite Cobalt Selenide Heterojunction and Its Electrocatalytic Performance

Tengfei Meng [1], Hongjin Shi [1], Feng Ao [1], Peng Wang [1], Longyao Wang [1], Lan Wang [1], Yujun Zhu [2], Yunxiang Lu [3] and Yupei Zhao [1,*]

[1] School of Petrochemical Engineering, Changzhou University, Changzhou 212006, China
[2] Department of Pharmaceutical and Biomedical Engineering, Clinical College of Anhui Medical University, Hefei 230031, China
[3] School of Chemical and Molecular Engineering, East China University of Science and Technology, Shanghai 200237, China
* Correspondence: zhaoyupei@cczu.edu.cn

Abstract: With the increasing utilization of clean energy, the development and utilization of hydrogen energy has become a research topic of great significance. Cobalt selenide (CS) is an electrocatalyst with great potential for oxygen evolution reaction (OER). In this paper, a nitrogen-doped biomass carbon (1NC@3)-based composite cobalt selenide (CS) heterojunction was prepared via a solvothermal method using kelp as the raw material. Structural, morphological, and electrochemical analyses were conducted to evaluate its performance. The electrochemical test results demonstrate that the overpotential of the CS/1NC@3 catalyst in the OER process was 292 mV, with a Tafel slope of 98.71 mV·dec^{-1} at a current density of 10 mA·cm^{-2}. The electrochemical performance of the CS/1NC@3 catalyst was further confirmed by theoretical calculations, which revealed that the presence of the biomass carbon substrate enhanced the charge transport speed of the OER process and promoted the OER process. This study provides a promising strategy for the development of efficient electrocatalysts for OER applications.

Keywords: cobalt selenide; electrocatalysis; biomass; DFT

1. Introduction

Traditional energy has brought severe environmental issues, and the depletion of non-renewable resources has prompted humans to develop and research new energy sources [1]. Hydrogen energy is an ideal clean energy source, and water electrolysis is an ideal process for producing hydrogen. Oxygen Evolution Reaction (OER), which takes place at the anode, is the rate-determining step of water electrolysis for hydrogen production and has become an area of extensive research [2–4]. Platinum (Pt) and its alloys have always been considered excellent electrocatalysts; however, their high cost and low durability significantly impede their commercial application. Biomass energy is an important energy source for human survival; thus, it is of great research significance to use biomass to replace non-renewable resources. The combination of the electrode material prepared from biomass as the carbon source with hydrogen energy is currently a research hotspot [5–7].

As a renewable resource, algae are widely found in oceans and lakes. Elements such as nitrogen, phosphorus, calcium, and iron are evenly distributed in biomass, which can be synthesized in situ into self-doped carbon materials after being carbonized without having to add additional heteroatom dopants [8]. The uniform distribution of elements can improve carrier mobility, thereby effectively improving the material's electrochemical performance. Pérez-Salcedo, K.Y et al. used Sargassum as a carbon source, activated it with potassium hydroxide, and then doped it with hydrazine sulfate to obtain a catalyst with a

surface area of 2289 m$^2 \cdot$g^{-1}, a nitrogen content of 0.16%, and a sulfur content of 2.63%. This showed good ORR activity [9]. Wang, F. et al. developed a self-doping electrocatalyst from rice husk biomass, where the self-doping of Si in the rice husk biomass improved the degree of graphitization of the catalyst. The prepared catalyst had good electrocatalytic activity, good stability (a retention rate greater than 85% after 40,000 s), and methanol tolerance [10]. Hao et al. used seaweed biomass sodium alginate to prepare defective carbon catalysts with good ORR activity and selectivity comparable to commercial Pt/C catalysts. The catalysts exhibited better stability and methanol tolerance than Pt/C catalysts [11]. It is clear that the research and development of biomaterials as electrocatalysts have the potential to replace noble metal catalysts.

Previous studies have shown that OER electrocatalysts based on transition metals typically demonstrate high over-potentials [12]. However, cobalt-based materials, including CoS_2 [13], CoSe [14], CoFe [15], and other related compounds, have garnered significant attention as a promising research area for low-cost and high-efficiency OER catalysts. Cobalt-based materials have unique electronic structures and abundant active sites. Currently, the methods to improve the OER performance of cobalt-based materials mainly include increasing the active sites by doping and adjusting the surface electronic structure or synergy by constructing heterostructures, thereby improving the electrocatalytic activity of the materials. Ke Zhang et al. doped $CoSe_2$ with iron by hydrothermal synthesis, and the resulting the $FeCoSe_2$/$Co_{0.85}$Se heterostructure catalyst exhibited good electrocatalytic activity at a current density of 10 mA·cm^{-2}. The overpotential under-density is 0.33 V, and the Tafel slope is 50.8 mV·dev^{-1} [16]. Yucan Dong et al. obtained the excellent OER performance of Co_9S_8@CoS_2 heterojunctions synthesized by hydrogen etching of CoS_2, and the Co_9S_8@CoS_2 heterostructures were dynamically reformed during OER with the in situ formation of CoOOH structure, thus exhibiting excellent electrocatalytic performance [17]. It can be seen that constructing heterostructures on cobalt-based materials is a common method to improve carrier mobility and thus improve electrocatalytic performance [18,19].

In recent years, we have witnessed the eutrophication of many water bodies due to pollution and biological invasions of the ecological environment. These factors cause algal blooms in some areas and severe ecological and environmental problems. On the other hand, the development and utilization of algae can turn waste into treasure and provide new ideas for developing and utilizing new energy. In this paper, kelp was used as the carbon precursor, and it was doped with nitrogen to design and synthesize the cobalt selenide composite heterojunction. The material's electrochemical test showed that the OER process' overpotential was 292 mV; the Tafel slope was 98.71 mV·dev^{-1}; and the electrochemical impedance was 125.95 Ω. The mechanism of the enhancement of the electrocatalytic activity by the heterostructure was further studied through theoretical calculations. These calculations confirm that the presence of the biomass carbon substrate enhances the charge transport of the OER process and promotes the OER process [20].

2. Experimental Section

2.1. Catalyst Preparation

The kelp was purchased from a local supermarket in Changzhou, China. Melamine ($C_3H_6N_6$), potassium hydroxide (KOH) and cobalt acetate tetrahydrate provided by Sinopharm Chemical Reagent Co., Ltd. (Shanghai, China), selenium powder (China Institute of Metal Materials), tube furnace (Anhui Jing Branch Co., Ltd. Hefei, China), and electrochemical workstation (CS350H, Wuhan Coster Co., Ltd. Wuhan, China) were used. Deionized water was used in all of the experiments.

Synthesis of NC: NC was synthesized following a combustion technique. First, 5 g of kelp was cut into pieces, washed, and mixed with KOH in a beaker. Ethanol was added to the beaker, and the mixture was then subjected to ultrasonic drying for 1 h, followed by vacuum-drying at 80 °C for 1 h. The mixture was transferred to a crucible and heated to 550 °C in a tube furnace at a heating rate of 5 °C·min^{-1}, kept warm for 2 h, and then washed to neutrality after cooling. A reddish powder was obtained. A certain amount of melamine was added to the reddish powder in a beaker, followed by the addition of ethanol to dissolve it. The mixture was ultrasonicated for 1 h and vacuum-dried at 80 °C. The mixture was heated to 900 °C in a tube furnace at a heating rate of 5 °C·min^{-1}, kept warm for 2 h, cooled, and dried to obtain a black powder. As a comparison, the amounts of melamine added were 6 g, 12 g, and 16 g, respectively, and the samples were marked as NC@X (X = 1, 2, 3). The schematic flow chart of this experiment is shown in Figure 1.

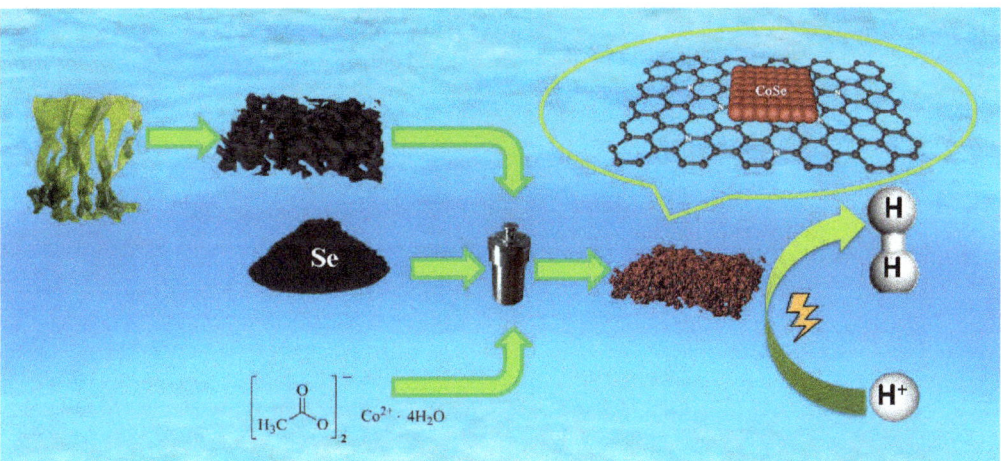

Figure 1. Schematic illustration of the synthesis of CS/yNC@X.

Preparation of CS/yNC@X: A nitrogen-doped biomass carbon-based composite cobalt selenide heterojunction was synthesized using a solvothermal method. An appropriate amount of NC@X, selenium powder, cobalt acetate tetrahydrate, sodium borohydride, and 23 mL of absolute ethanol was placed in a 50 mL hydrothermal kettle and kept at 180 °C for 18 h. After cooling, the sample was rinsed with deionized water and absolute ethanol, dried, and finally obtained as a black powder. For comparison, samples with 0.05 g, 0.1 g, and 0.2 g of NC@X added were labeled as CS/yNC@X (y = 1, 2, 3).

Preparation of CoSe: Cobalt selenide was combined using the solvothermal method. First, we took an appropriate amount of selenium powder, cobalt acetate tetrahydrate, sodium borohydride, and 23 mL of absolute ethanol. We put them in a 50 mL hydrothermal kettle to keep warm at 180 °C for 18 h, then washed them with deionized water and absolute ethanol after cooling down. Finally, they were dried to obtain a black powder.

2.2. Theoretical Calculations

All calculations were performed using the Dmol3 module from Materials Studio 2020. The GGA/PBE functional was used in the DFT simulation, and the double value plus polarization basis set was used. The isoelectronic calculation of the complex showed that the spin state of the system had no spin limitation, and the SCF self-consistent convergence accuracy was 1.0×10^{-6}. The allowable deviations for the total energy, gradient, and displacement were 1.0×10^5 Ha, 0.002 Ha·Å^{-1}, and 0.005 Å, respectively. In order to avoid the influence of Brillouin zone sampling, the k point was set to $3 \times 3 \times 1$ to ensure the rationality of the calculation results.

2.3. Physical Characterization

The crystal phase of the catalyst was characterized by X-ray. Its morphology was characterized by scanning electron (Nova450) and transmission electron (TEM) (Jem2100 F), and its chemical composition and elemental valence states were determined by X-ray photoelectron spectroscopy (Brooke D8 advance).

2.4. Electrochemical Measurements

The electrochemical performance of the prepared product was tested using the method of manufacturing powder electrodes. First, the reference electrode was converted to a standard hydrogen electrode (SHE) using the conversion formula:

$$E\{RHE\} = E\{SCE\} + 0.243\ V + 0.0591 \times pH \tag{1}$$

Next, a uniformly dispersed coating solution was obtained by dispersing 2 mg of catalyst in 360 µL isopropanol, 120 µL water, and 20 µL Nafion (5 wt%) and then sonicating the mixture for 20 min. Then, 12 µL of the obtained solution was coated on a 3 mm glassy carbon electrode and dried, forming the desired powder electrode. The electrochemical test was carried out using an electrochemical workstation (CS350H, Wuhan Kesite Instrument Co., Ltd., Wuhan, China) and a 1 M KOH solution (pH = 14). The volt–ampere characteristic curve was scanned at a speed of 5 mV·s^{-1} and the electrochemical impedance spectroscopy test was carried out under 5 mV AC voltage, with the scanning frequency ranging from 0.1 Hz to 105 Hz. To determine the stability of the catalyst, the voltage of −0.5 V was scanned for 14 h to record the current change.

3. Results

The microstructures of CS/1NC@X were investigated using SEM. Figure 2 showed that CS/1NC@X had a distinct three-dimensional coral-like structure. The three-dimensional coral-like structure of CS/1NC@X in the figure is composed of cobalt selenide particles tens of nanometers in size. Cobalt selenide is complexed with NC@X through various forms such as intercalation, loading, and encapsulation, among which the loading type is dominant. In the overall structure, NC@X acts as a carrier for CoSe loading and fixes CoSe into a closely packed structure.

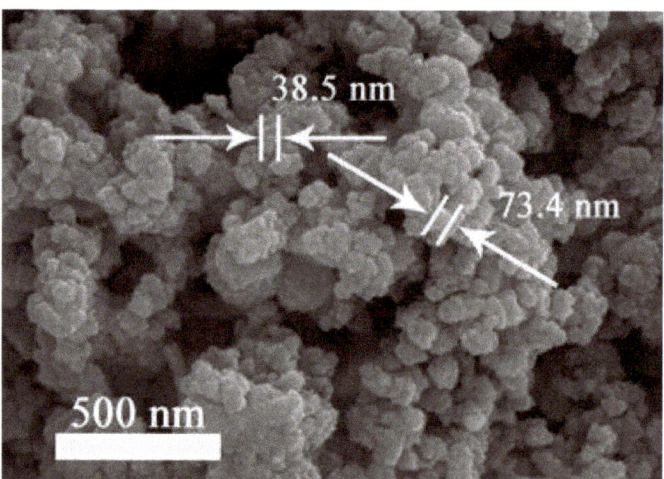

Figure 2. SEM images of CS/1NC@3.

From the TEM and HRTEM results of CS/1NC@3 (Figure 3a,b), it can be clearly observed that the nano-substrate structure of NC is uniformly loaded with CoSe nanoparticles to form a heterostructure. This helps to facilitate quick electron transfer from NC@X to CoSe, thereby improving the material's conductivity and surface binding energy and accelerating the electrocatalytic process. In high-resolution TEM (HRTEM) images, lattice spacings of 0.323 nm, 0.271 nm, and 0.218 nm were determined, corresponding to the CoSe (1 0 0), (1 0 1), and (1 0 2) crystal planes, respectively. The interlayer spacing is close to the ideal spacing, indicating that the degree of graphitization of the NC substrate after firing is higher, and the conductivity is enhanced [21].

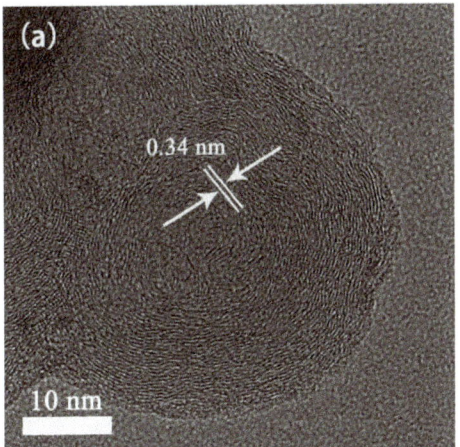 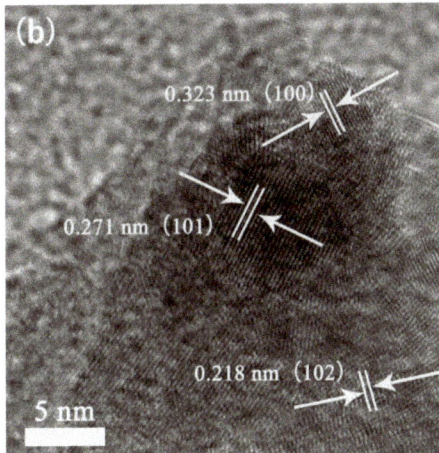

Figure 3. TEM image (**a**) and HRTEM image (**b**) of CS/1NC@3.

The results of X-ray powder diffraction (Figure 4) show that there are three strong diffraction peaks in Figure 4, which correspond to the (1 0 0) plane, (1 0 1) plane, and (1 0 2) plane of CoSe, respectively. Specifically, the (1 0 1) plane shows a strong diffraction peak intensity, which is attributed to the fact that it is the active plane of CoSe. In addition, the NC@X diffraction peak appears on CoSe, and the (1 0 0) peak corresponding to the carbon material shifts to the left. This indicates that after N doping, the increase in the interlayer spacing leads to the change in the energy band filling state and the Fermi energy level, which enhances the conductivity, ion transport ability, and catalytic activity of the material. The increase in the interlayer spacing can expose more active sites and unsaturated defect sites, which can effectively improve the electrocatalytic performance [22–24]. Calculated by Scherrer's formula, the average grain size of CoSe was 19.3 nm and the average grain size of CS/1NC@3 was 28.3 nm. This shows that CoSe and NC@3 form a heterostructure and promote the formation of CS/1NC@3 phase.

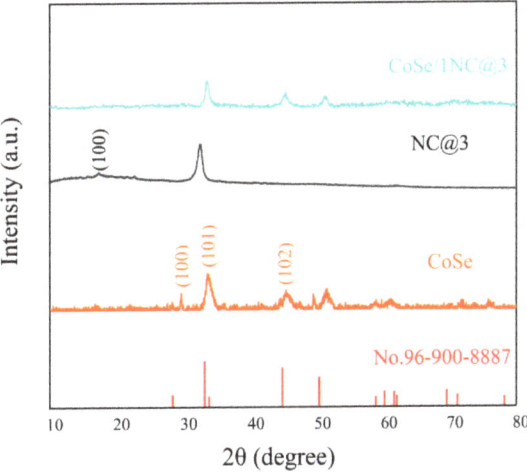

Figure 4. XRD pattern of CS/1NC@3, CoSe, NC@3.

The chemical states and bonding structures of CS/1NC@3 and their components were investigated by X-ray Photoelectron Spectroscopy (XPS) (Figure 5a–d). The high-resolution C 1s spectrum of CoSe/NC can be decomposed into peaks: C-C (284.8 eV), N-C (290.8 eV) and C-O (286.4 eV). Nitrogen is doped into the skeleton of kelp carbon by co-calcination, with a N-to-C ratio of approximately 1:6. Since nitrogen atoms are more electronegative than carbon atoms, nitrogen-doped carbon materials will have better electronic conductivity. After nitrogen atoms are doped into the carbon material framework, they can provide more free electrons for the conduction band, significantly improving the electrical conductivity of the carbon framework [25]. N 1s can be decomposed into two peaks at 397.8 eV and 405.7 eV, respectively, for the pyridinic nitrogen and N-O structures. These are related to the small amount of oxygen contained in kelp itself, which may be partially oxidized during the calcination process. In addition, Co 2p is affected by orbital hybridization and can be decomposed into Co 2p3/2 and Co 2p1/2. The valence state of the element is mainly +2. Compared with pure CoSe, the binding energy shifts to the low energy direction by about 0.25 eV, and the binding energy of the C-C bond can also shift to the high-energy direction by about 0.2 eV, which shows that in the CoSe/NC heterojunction, there is a strong electronic coupling between the CN structure of the substrate and CoSe. The charge transport direction is CN-Co. The CN structure of the substrate injects a small number of electrons into CoSe, thus improving its activity [26,27]. Se is decomposed into two electron orbitals, Se 3d5/2 (54.4 eV) and Se 3d3/2 (60.1 eV), which exist in the form of CoSe conjugates.

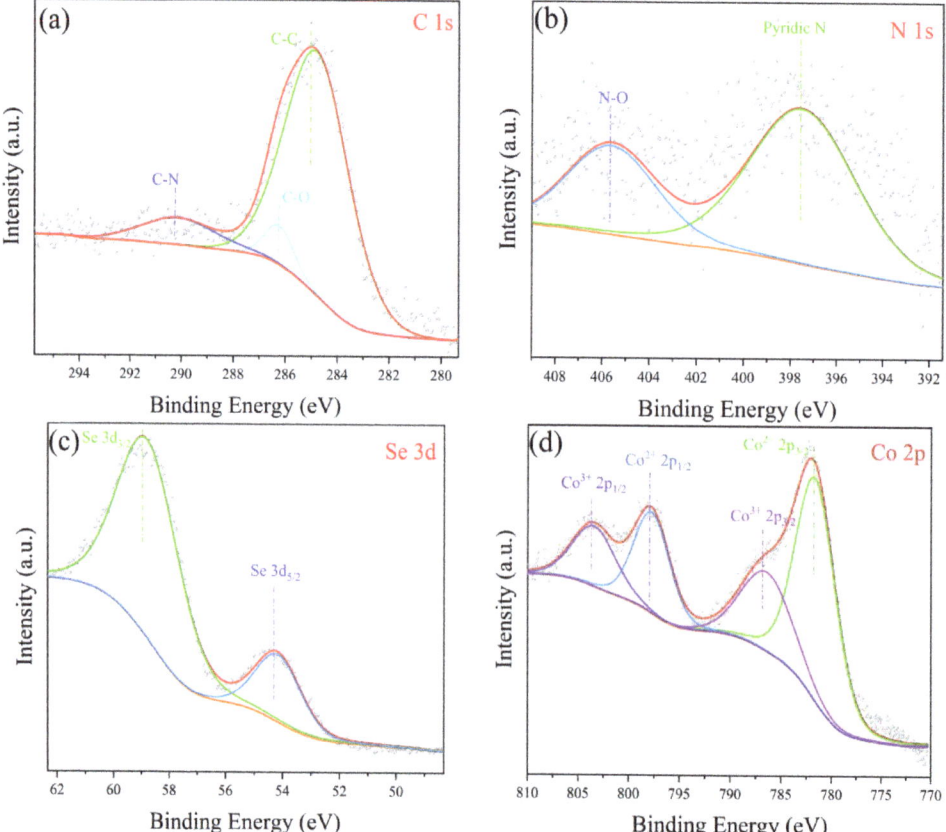

Figure 5. XPS spectra of C 1s (**a**), N 1s (**b**), Se 3d (**c**), and Co 2p (**d**) spectra for CS/1NC@3.

The presence of multiple oxygen-containing functional groups on the surface of carbon materials can promote the strong adsorption of metal particles, resulting in an improved dispersion of the metal particles [4]. The CS/1NC@3 was further investigated by Fourier transform infrared (FTIR) spectroscopy. The FTIR spectrum of CS/1NC@3 exhibited peaks at 3380 cm^{-1}, 2925 cm^{-1}, 1627 cm^{-1}, 1385 cm^{-1}, and 1045 cm^{-1}, which were attributed to the -OH, C-H, C=N, C=O and C=N functional groups, respectively (Figure 6). The infrared spectrum showed that the oxygen-containing functional groups on the material may be phenolic hydroxyl, carboxyl, lactone, or carbonyl [28]. However, the presence of a small amount of oxygen-containing groups can increase surface wetting reactivity and electrochemical performance.

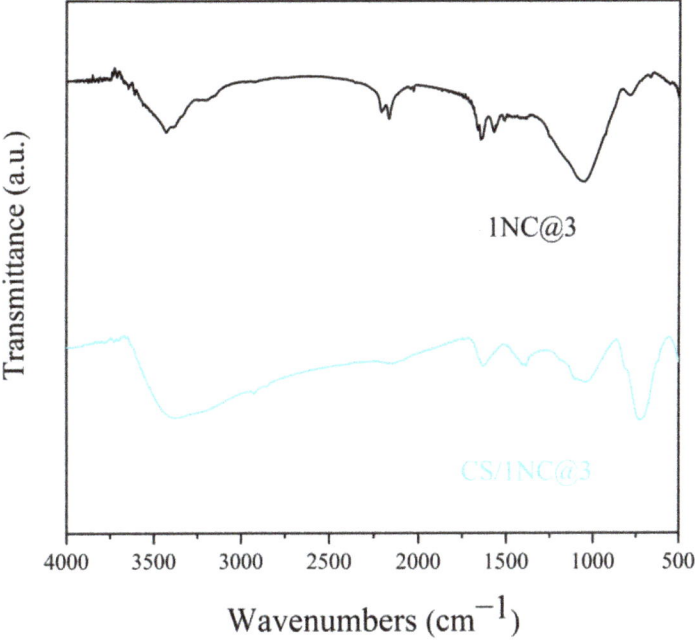

Figure 6. FTIR spectra of the CS/1NC@3 and NC@3.

The electrochemical performance test results of CS/1NC@3 materials are shown in Figure 7a–c. As can be seen from the linear sweep voltammetry (LSV) curve, the overpotential of CS/1NC@3 is the lowest, which is 292.7 mV at a current density of 10 mA·cm^{-2}, while that of pure CoSe is 384.8 mV. This demonstrates that CoSe and N-doped biochar have formed a heterostructure, which increases the charge carrier mobility and exhibits superior electrocatalytic activity compared to the original CoSe [29]. At the same time, the influence of the NC substrate-modified CoSe electrode on the kinetics of the electrocatalytic OER reaction was analyzed by the Tafel diagram, in which the Tafel slope (98.71 mV·dec^{-1}) of CS/1 NC@3 was smaller than that of CoSe (258.31 mV·dec^{-1}). The decrease in the Tafel slope indicates that the addition of NC changes the rate-determining step but plays a certain role in enhancing the kinetic process of the electrochemical OER. To further explore the reason for the enhanced activity, the EIS of the material was tested from the perspective of charge transfer kinetics. The results of electrochemical impedance analysis showed that, compared with CoSe (300.54 Ω), CS/1NC@3 (125.95 Ω) composites had smaller charge transfer resistance and thus higher activity [30].

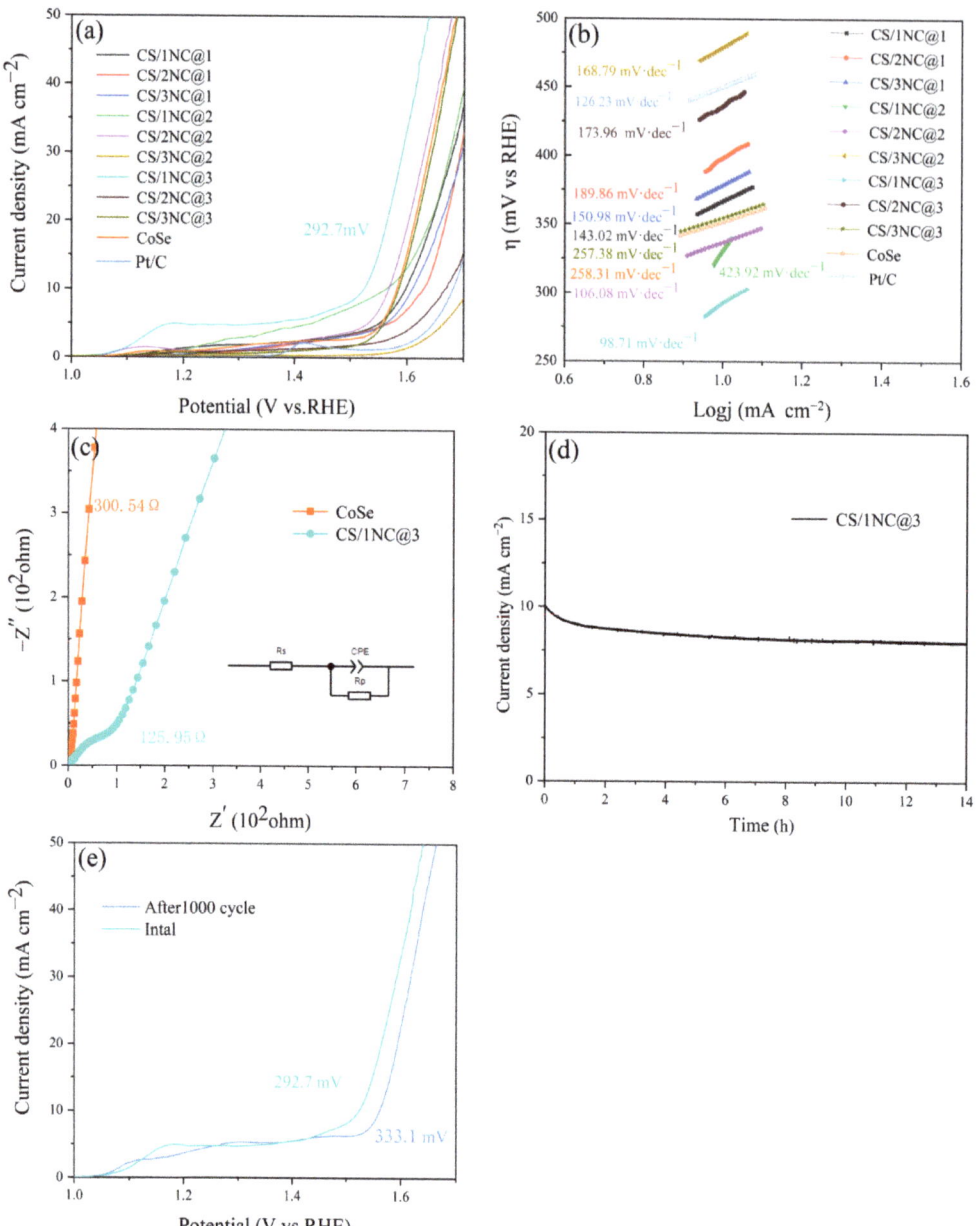

Figure 7. (**a**) LSV curve (without iR-corrected) and (**b**) Tafel slope of CS/yNC@X, (**c**) Nyquist plots of CS/1NC@3 and CoSe, (**d**) i–t curves of CS/1NC@3. (**e**) LSV curves of CS/1NC@3 in 1 M KOH solutions, respectively, before and after 1000 cycles.

The CS/1NC@3 catalyst was kept in 1 M KOH for 14 h in Figure 7d, which indicated that the heterojunction formed between cobalt selenide and biochar material by the hydrothermal method had good stability under alkaline conditions. In Figure 7e, we examined the stability of the catalytic electrode CS/1NC@3 LSV curves after 1000 consecutive CV cycle values remained at 333.1 mV, which further confirmed the high stability of CS/1NC@3. This high stability can be attributed to the strong bond between NC@3 and CoSe.

DFT calculation: Generally, a good electrocatalyst has excellent charge capture and transfer capabilities; the key lies in the electronic structure and conductivity of the catalyst. Based on the XPS surface element analysis results, a heterostructure model was constructed using Materials Studio. The (0 0 1) surface of CoSe was chosen as the active surface, and a 2 × 2 supercell was created in the X and Y directions. The lattice lengths of CoSe after the supercell expansion in the X and Y directions were 7.21 Å, while that of CN was 7.38 Å. The overall lattice mismatch degree after constructing the heterojunction was less than 5%, which is considered reasonable. Based on the density of states (DOS) diagrams of CS/yNC@X and the CoSe catalyst depicted in Figure 8a, b, respectively, the d-band theory suggests that when the d-band center of the transition metal is positioned closer to the Fermi level, the anti-bonding orbital located above the Fermi level becomes elevated after the catalyst and the adsorbate orbital form a bond. As a result, there is a reduced probability of electron filling in the anti-bonding orbital, which leads to more stable adsorption, stronger adsorption energy, and enhanced catalytic activity [31]. From the PDOS plot of CS/yNC@X, it can be observed that the heterostructure still exhibits strong metallic properties after doping with nitrogen. The enhanced metallic properties of the biomass-based carbon matrix after nitrogen doping indicate that this doping method can improve the system's conductivity, which is beneficial for promoting electrochemical reactions. The d-band center of pure CoSe is at 2.14 eV, and the CS/1NC@3 d-band center is at 2.093 eV after projecting the polarized density of the state of CS/1NC@3 to the Fermi surface. The center is obviously shifted to the Fermi level. It is clear that the activity of the CoSe material is significantly enhanced by the method of constructing a heterojunction, and the OER reaction is better promoted, which is consistent with the experimental results.

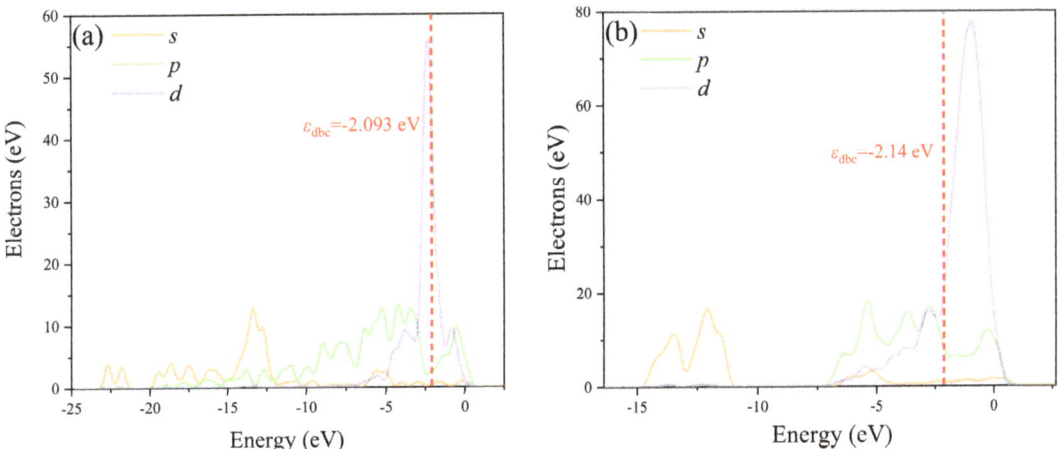

Figure 8. (a) CS/yNC@X density of states diagram, (b) CoSe density of states diagram.

Through the calculation of the differential charge of the material, a section perpendicular to the heterojunction material is constructed along the A B direction, and the heterostructure model constructed by Materials Studio and the section diagram of the average differential charge density of the material along the Z direction are obtained. Figure 9b, c. It is clear that the direction of charge transfer is from the carbon-nitrogen substrate to CoSe, and the charge is concentrated at the Co atom (the red part in the figure). Co has a higher catalytic activity after receiving electrons, which is consistent with the current general conclusions. Gaining electrons promotes the reactivity of metals [32–34]. The electrostatic potential analysis also shows that the doping of nitrogen improves the conductivity of the carbon-based substrate and that there is a charge transfer from C to N inside the carbon–nitrogen substrate.

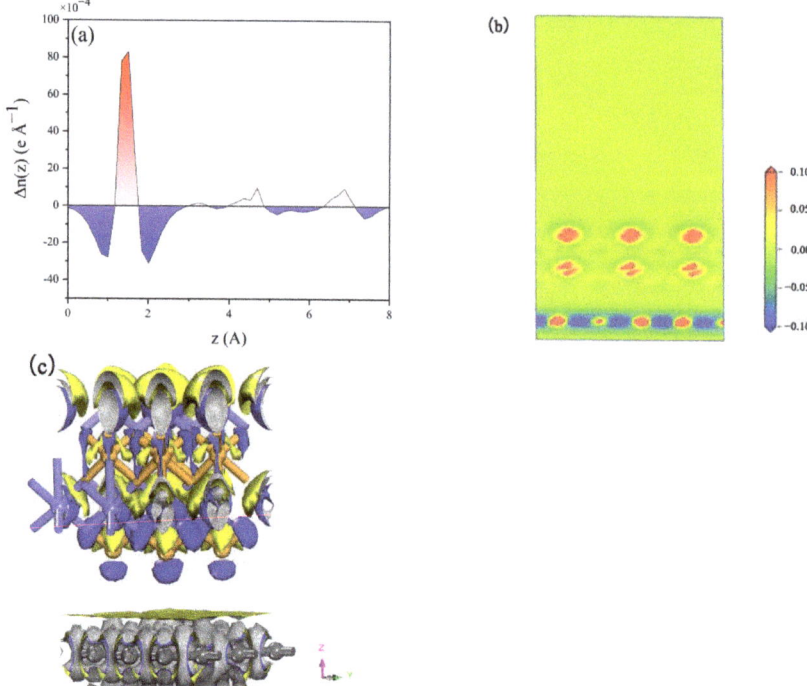

Figure 9. (**a**) CS/yNC@X differential charge density map, (**b**) CS/yNC@X differential charge section map, (**c**) heterostructure electrostatic potential distribution map.

The four-electron process is Nørskov's classical theory, and this process is currently generally accepted in academia [35,36]. According to the four-electron process of OER [37,38], the decomposition of water on the catalyst proceeds in four steps:

$$Catalyst + H_2O = {}^*OH + H^+ + e^- \tag{2}$$

$$^*OH = {}^*O + H^+ + e^- \tag{3}$$

$$^*O + H_2O = {}^*OOH + H^+ + e^- \tag{4}$$

$$^*OOH = * + O_2 + H^+ + e^- \tag{5}$$

In order to calculate the energy of each step correctly, the initial adsorption configuration was optimized, as shown in Figure 10a–d. There were three possible adsorption active centers on CoSe: Co inside, Se, and Co on edge. These three sites were used as active centers to calculate the energy and configuration changes of the adsorbed OH. After the adsorption of OH by marginal Co, the overall energy is the lowest, which is −219.18 eV), making it the most likely site at which OER will occur. After the adsorption of OH by internal Co, the overall energy is −219.03 eV. In comparison, the overall energy after the adsorption of OH by Se as the active center is −218.12 eV, which is the highest amount of energy. It is difficult for OH adsorption to occur at the Se site. By analyzing the bond length between the three types of active centers and O after the adsorption of OH, it can be found that the adsorption energy of OH adsorbed by the edge Co is more negative, so the distance between Co-O is closer, which is 1.8 Å. After the adsorption of OH by the internal Co, the Co-O distance is 2.08 Å. After the adsorption of OH by Se, the OH tends to be far away from Se and close to Co after structural optimization. It is difficult for surface OH to adsorb on the Se surface, and the Se-O distance of 3.14 Å also indicates that Se is not the active center of the OER process. The adsorption energy of marginal Co is greater than that of internal Co, which may be due to the exposure of more coordination sites of marginal Co. Therefore, the calculation of the four-electron process step diagram uses marginal Co as the active site to expand the calculation.

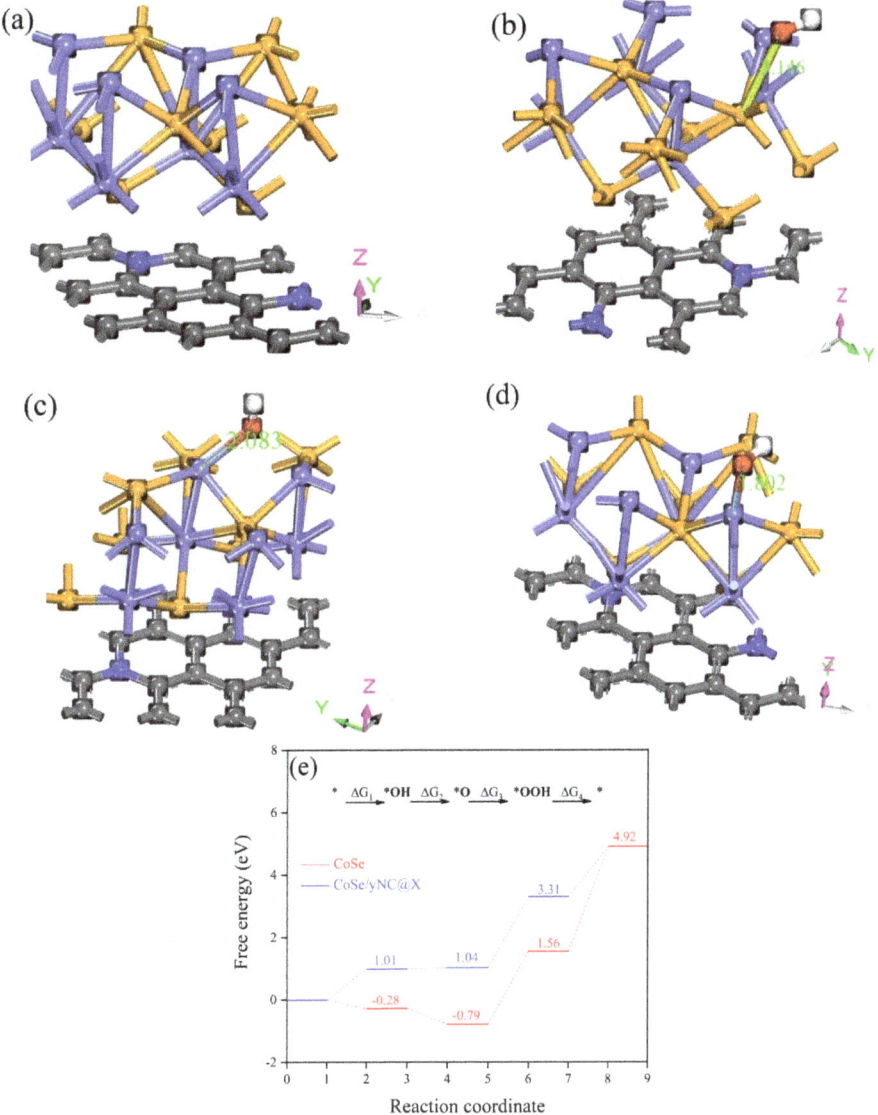

Figure 10. (**a**–**d**) are four simulated configurations, (**e**) OER process step diagram of CoSe and CoSe/yNC@X.

The free energy of each reaction step of pure CoSe and CS/yNC@X catalyst OER was calculated, as shown in Figure 10e. The change of free energy evaluated the intrinsic activity of the catalyst, and the change of free energy of pure CoSe and the CS/yNC@X catalyst OER process were compared. The free energy of the second and third steps of pure CoSe changes greatly, which seriously affects the reaction rate. *OH is converted to *O in the second step of the final speed step, and the energy barrier is as high as 3.25 eV, which makes it difficult to carry out the kinetic process of water electrolysis catalyzed by pure CoSe. The rate-determining step of the CS/yNC@X heterojunction is the formation of the second step, OOH intermediate, with an energy barrier of 1.42 eV. Therefore, by synthesizing heterojunction materials, the kinetic process of the catalyst surface reaction is changed, the velocity step is changed, and the energy barrier of the rate-determining step is effectively reduced, thus improving the performance of OER.

4. Conclusions

Utilizing Marine biomass, through the two-step carbonization and solvothermal synthesis methods, with selenium powder as the selenium source, cobalt acetate tetrahydrate as the cobalt source, kelp as the carbon precursor, KOH activation, and using melamine as the nitrogen source, carbon/heterojunction composites containing rich nitrogen doping were prepared. The strong interfacial interaction of the two components can establish abundant high-speed electron transmission channels. The synthesized material has CS/1NC@3 biomass porous carbon structure and abundant active sites. Through XPS analysis and other characterization methods, the element composition of the surface and the electron transfer method were judged, and the synthesized material was finally confirmed as CS/1NC@3. According to the electrochemical performance test and characterization, under the current density of 10 mA·cm^{-2}, the required overpotential is 292.7 mV, and the Tafel slope is 98.71 mV·dev^{-1}. Compared with CoSe alone, the center of the D-band shifts towards the Fermi surface, demonstrating enhanced catalytic activity. According to the results of the step diagram of the four-electron process, by directly doping nitrogen with carbon substrate, the overpotential of the process was significantly reduced, and the OER process was promoted. CS/1NC@3 showed significantly enhanced OER activity, which was consistent with the theoretical settlement results.

Author Contributions: L.W. (Longyao Wang), L.W. (Lan Wang); Y.Z. (Yupei Zhao) and P.W. planned and supervised this work; T.M. and F.A. prepared the multilayer films and co-wrote and edited this paper with Y.Z. (Yupei Zhao); Y.Z. (Yupei Zhao) and P.W. supervised the Masters student T.M. and F.A. performed some experiments and tests; T.M. and H.S. performed and analyzed characterization of materials and samples; Y.Z. (Yujun Zhu) and Y.L. provided simulations and calculations. All authors have read and agreed to the published version of the manuscript.

Funding: National Natural Science Foundation of China, No. 22072007. The Jiangsu Province Project of Industry-University-Research Cooperation, No. BY20221141. Jiangsu Graduate Research and Practice Innovation Program, No. SJCX21_1252. Jiangsu College Student Innovation and Entrepreneurship Training Program, No. 202210292097Y. Zhenjiang Key Research and Development Program (Industry Foresight and Common Key Technologies), No. CQ2022006. Jiangsu Engineering Technology Research Center for Novel Anti-Influenza Virus Drugs, Jiangsu Graduate Workstation, Zhenjiang Salicylic Acid Series Products Engineering Technology Research Center, Zhenjiang Municipal Enterprise Technology Center.

Institutional Review Board Statement: Not applicable.

Informed Consent Statement: Not applicable.

Data Availability Statement: Not applicable.

Acknowledgments: The author is immensely grateful to Shuo Li for his assistance with English writing and Chengdong Wang for his help in the experiment.

Conflicts of Interest: The authors declare no conflict of interest.

References

1. Gao, Y.; Zhang, N.; Wang, C.; Zhao, F.; Yu, Y. Construction of Fe_2O_3@CuO Heterojunction Nanotubes for Ellen Vayner hanced Oxygen Evolution Reaction. *ACS Appl. Energy Mater.* **2019**, *3*, 666–674. [CrossRef]
2. Basu, M.; Zhang, Z.W.; Chen, C.J.; Lu, T.H.; Hu, S.F.; Liu, R.S. $CoSe_2$ Embedded in C_3N_4: An Efficient Photocathode for Photoelectrochemical Water Splitting. *ACS Appl. Mater. Inter.* **2016**, *8*, 26690–26696. [CrossRef] [PubMed]
3. Campos-Roldán, C.A.; Alonso-Vante, N. The Oxygen Reduction and Hydrogen Evolution Reactions on Carbon Supported Cobalt Diselenide Nanostructures. *J. Electrochem. Soc.* **2020**, *167*, 026507. [CrossRef]
4. Ye, Y.-Y.; Qian, T.-T.; Jiang, H. Co-Loaded N-Doped Biochar as a High-Performance Oxygen Reduction Reaction Electrocatalyst by Combined Pyrolysis of Biomass. *Ind. Engineering. Chem. Res.* **2020**, *59*, 15614–15623.
5. Ma, L.L.; Hu, X.; Liu, W.J.; Li, H.C.; Lam, P.K.S.; Zeng, R.J.; Yu, H.Q. Constructing N, P-dually doped biochar materials from biomass wastes for high-performance bifunctional oxygen electrocatalysts. *Chemosphere.* **2021**, *278*, 130508. [CrossRef]
6. Ma, N.; Jia, Y.; Yang, X.; She, X.; Zhang, L.; Peng, Z.; Yao, X.; Yang, D. Seaweed biomass derived (Ni,Co)/CNT nanoaerogels: Efficient bifunctional electrocatalysts for oxygen evolution and reduction reactions. *J. Mater. Chem. A* **2016**, *4*, 6376–6384. [CrossRef]
7. Sekar, S.; Sim, D.H.; Lee, S. Excellent Electrocatalytic Hydrogen Evolution Reaction Performances of Partially Graphitized Activated-Carbon Nanobundles Derived from Biomass Human Hair Wastes. *Nanomaterials.* **2022**, *12*, 531. [CrossRef]
8. Huang, N.-B.; Zhang, J.-J.; Sun, Y.; Sun, X.-N.; Qiu, Z.-Y.; Ge, X.-W. A non-traditional biomass-derived N, P, and S ternary self-doped 3D multichannel carbon ORR electrocatalyst. *New J. Chem.* **2020**, *44*, 14604–14614. [CrossRef]
9. Pérez-Salcedo, K.Y.; Alonso-Lemus, I.L.; Quintana, P.; Mena-Durán, C.J.; Barbosa, R.; Escobar, B. Self-doped Sargassum spp. derived biocarbon as electrocatalysts for ORR in alkaline media. *Int. J. Hydrogen Energy.* **2019**, *44*, 12399–12408. [CrossRef]
10. Wang, F.; Li, Q.; Xiao, Z.; Jiang, B.; Ren, J.; Jin, Z.; Tang, Y.; Chen, Y.; Li, X. Conversion of rice husk biomass into electrocatalyst for oxygen reduction reaction in Zn-air battery: Effect of self-doped Si on performance. *J. Colloid Interface Sci.* **2022**, *606*, 1014–1023. [CrossRef]
11. Hao, Y.; Zhang, X.; Yang, Q.; Chen, K.; Guo, J.; Zhou, D.; Feng, L.; Slanina, Z. Highly porous defective carbons derived from seaweed biomass as efficient electrocatalysts for oxygen reduction in both alkaline and acidic media. *Carbon* **2018**, *137*, 93–103. [CrossRef]
12. Xu, J.; Li, J.; Lian, Z.; Araujo, A.; Li, Y.; Wei, B.; Yu, Z.; Bondarchuk, O.; Amorim, I.; Tileli, V.; et al. Atomic-Step Enriched Ruthenium–Iridium Nanocrystals Anchored Homogeneously on MOF-Derived Support for Efficient and Stable Oxygen Evolution in Acidic and Neutral Media. *ACS Catal.* **2021**, *11*, 3402–3413. [CrossRef]
13. Zhan, Y.; Yu, S.Z.; Luo, S.H.; Feng, J.; Wang, Q. Nitrogen-Coordinated CoS_2@NC Yolk-Shell Polyhedrons Catalysts Derived from a Metal-Organic Framework for a Highly Reversible $Li-O_2$ Battery. *ACS Appl. Mater. Inter.* **2021**, *13*, 17658–17667. [CrossRef] [PubMed]
14. Sobhani, A.; Salavati-Niasari, M. Cobalt selenide nanostructures: Hydrothermal synthesis, considering the magnetic property and effect of the different synthesis conditions. *J. Mol. Liq.* **2016**, *219*, 1089–1094. [CrossRef]
15. Dang, N.K.; Tiwari, J.N.; Sultan, S.; Meena, A.; Kim, K.S. Multi-site catalyst derived from Cr atoms-substituted CoFe nanoparticles for high-performance oxygen evolution activity. *Chem. Eng. J.* **2021**, *404*, 12653. [CrossRef]
16. Zhang, K.; Shi, M.; Wu, Y.; Wang, C. Constructing $FeCoSe_2$/$Co_{0.85}Se$ heterostructure catalysts for efficient oxygen evolution. *J. Alloys Compd.* **2020**, *825*, 154073. [CrossRef]
17. Dong, Y.; Ran, J.; Liu, Q.; Zhang, G.; Jiang, X.; Gao, D. Hydrogen-etched CoS_2 to produce a Co_9S_8@CoS_2 heterostructure electrocatalyst for highly efficient oxygen evolution reaction. *RSC Adv.* **2021**, *11*, 30448–30454. [CrossRef] [PubMed]
18. Tang, Y.; Jing, F.; Xu, Z.; Zhang, F.; Mai, Y.; Wu, D. Highly Crumpled Hybrids of Nitrogen/Sulfur Dual-Doped Graphene and Co_9S_8 Nanoplates as Efficient Bifunctional Oxygen Electrocatalysts. *ACS Appl. Mater. Interfaces.* **2017**, *9*, 12340–12347. [CrossRef]
19. Feizi, H.; Bagheri, R.; Song, Z.; Shen, J.-R.; Allakhverdiev, S.I.; Najafpour, M.M. Cobalt/Cobalt Oxide Surface for Water Oxidation. *ACS Sustain. Chem. Eng.* **2019**, *7*, 6093–6105. [CrossRef]
20. Wu, Y.; Wang, F.; Ke, N.; Dong, B.; Huang, A.; Tan, C.; Yin, L.; Xu, X.; Hao, L.; Xian, Y.; et al. Self-supported cobalt/cobalt selenide heterojunction for highly efficient overall water splitting. *J. Alloys Compd.* **2022**, *925*, 166683. [CrossRef]
21. Lian, P.; Zhu, X.; Liang, S.; Li, Z.; Yang, W.; Wang, H. Large reversible capacity of high quality graphene sheets as an anode material for lithium-ion batteries. *Electrochim. Acta.* **2010**, *55*, 3909–3914. [CrossRef]
22. Liang, H.; Jia, L.; Chen, F.; Jing, S.; Tsiakaras, P. A novel efficient electrocatalyst for oxygen reduction and oxygen evolution reaction in $Li-O_2$ batteries: Co/CoSe embedded N, Se co-doped carbon. *Appl. Catal. B* **2022**, *317*, 121698. [CrossRef]
23. Zhang, T.; Yu, J.; Guo, H.; Liu, J.; Liu, Q.; Song, D.; Chen, R.; Li, R.; Liu, P.; Wang, J. Heterogeneous $CoSe_2$–CoO nanoparticles immobilized into N-doped carbon fibers for efficient overall water splitting. *Electrochim. Acta* **2020**, *356*, 136822. [CrossRef]
24. Xue, Y.; Zhang, Q.; Wang, W.; Cao, H.; Yang, Q.; Fu, L. Opening Two-Dimensional Materials for Energy Conversion and Storage: A Concept. *Adv. Energy Mater.* **2017**, *7*, 1602684. [CrossRef]
25. Sheng, Z.; Shao, L.; Chen, J.; Bao, W.; Wang, F.; Xia, X. Catalyst-Free Synthesis of Nitrogen-Doped Graphene via Thermal Annealing Graphite Oxide with Melamine and Its Excellent Electrocatalysis. *ACS Nano.* **2011**, *5*, 4350–4358. [CrossRef]
26. Sam, D.K.; Sam, E.K.; Lv, X. Application of Biomass-Derived Nitrogen-Doped Carbon Aerogels in Electrocatalysis and Supercapacitors. *ChemElectroChem.* **2020**, *7*, 3695–3712.

27. Kumar, S.; Jena, A.; Hu, Y.C.; Liang, C.; Zhou, W.; Hung, T.F.; Chang, W.S.; Chang, H.; Liu, R.S. Cobalt Diselenide Nanorods Grafted on Graphitic Carbon Nitride: A Synergistic Catalyst for Oxygen Reactions in Rechargeable Li-O_2 Batteries. *ChemElectroChem.* **2018**, *5*, 29–35. [CrossRef]
28. Hei, Y.; Li, X.; Zhou, X.; Liu, J.; Sun, M.; Sha, T.; Xu, C.; Xue, W.; Bo, X.; Zhou, M. Electrochemical sensing platform based on kelp-derived hierarchical meso-macroporous carbons. *Anal. Chim. Acta* **2018**, *1003*, 16–25. [CrossRef]
29. Zhou, Y.; Tian, R.; Duan, H.; Wang, K.; Guo, Y.; Li, H.; Liu, H. CoSe/Co nanoparticles wrapped by in situ grown N-doped graphitic carbon nanosheets as anode material for advanced lithium ion batteries. *J. Power Sources.* **2018**, *399*, 223–230. [CrossRef]
30. Jiang, D.; Xu, Q.; Meng, S.; Xia, C.; Chen, M. Construction of cobalt sulfide/graphitic carbon nitride hybrid nanosheet composites for high performance supercapacitor electrodes. *J. Alloys Compd.* **2017**, *706*, 41–47. [CrossRef]
31. Hammer, B.; Nørskov, J. Electronic factors determining the reactivity of metal surfaces. *Surf. Sci.* **1995**, *343*, 211–220. [CrossRef]
32. Vayner, E.; Sidik, R.A.; Anderson, A.B.; Popov, B.N. Experimental and Theoretical Study of Cobalt Selenide as a Catalyst for O_2 Electroreduction. *J. Phys. Chem. C* **2007**, *111*, 10508–10513. [CrossRef]
33. Wang, X.; Zhuang, L.; He, T.; Jia, Y.; Zhang, L.; Yan, X.; Gao, M.; Du, A.; Zhu, Z.; Yao, X.; et al. Grafting Cobalt Diselenide on Defective Graphene for Enhanced Oxygen Evolution Reaction. *iScience.* **2018**, *7*, 145–153. [CrossRef] [PubMed]
34. Li, K.; Cheng, R.; Xue, Q.; Meng, P.; Zhao, T.; Jiang, M.; Guo, M.; Li, H.; Fu, C. In-situ construction of Co/CoSe Schottky heterojunction with interfacial electron redistribution to facilitate oxygen electrocatalysis bifunctionality for zinc-air batteries. *Chem. Eng. J.* **2022**, *450*, 137991. [CrossRef]
35. Harzandi, A.M.; Shadman, S.; Nissimagoudar, A.S.; Kim, D.Y.; Lim, H.D.; Lee, J.H.; Kim, M.G.; Jeong, H.Y.; Kim, Y.; Kim, K.S. Ruthenium Core–Shell Engineering with Nickel Single Atoms for Selective Oxygen Evolution via Nondestructive Mechanism. *Adv. Energy Mater.* **2021**, *11*, 2003448. [CrossRef]
36. Medford, A.J.; Moses, P.G.; Jacobsen, K.W.; Peterson, A.A. A Career in Catalysis: Jens Kehlet Nørskov. *ACS Catal.* **2022**, *12*, 9679–9689. [CrossRef]
37. Liu, P.; Yan, J.Y.; Mao, J.X.; Li, J.W.; Liang, D.X.; Song, W.B. In-plane intergrowth CoS_2/MoS_2 nanosheets: Binary metal–organic framework evolution and efficient alkaline HER electrocatalysis. *J. Mater. Chem. A* **2020**, *8*, 11435. [CrossRef]
38. Nørskov, J.K.; Rossmeisl, J.; Logadottir, A.; Lindqvist, L.; Kitchin, J.R.; Bligaard, T.; Jónsson, H. Origin of the Overpotential for Oxygen Reduction at a Fuel-Cell Cathode. *J. Phys. Chem. B* **2004**, *108*, 17886–17892. [CrossRef]

Disclaimer/Publisher's Note: The statements, opinions and data contained in all publications are solely those of the individual author(s) and contributor(s) and not of MDPI and/or the editor(s). MDPI and/or the editor(s) disclaim responsibility for any injury to people or property resulting from any ideas, methods, instructions or products referred to in the content.

Article

A Determination of the Influence of Technological Parameters on the Quality of the Created Layer in the Process of Cataphoretic Coating

Jozef Dobránsky [1,*], Miroslav Gombár [2], Patrik Fejko [1] and Róbert Balint Bali [1]

[1] Faculty of Manufacturing Technologies with a Seat in Presov, Technical University of Kosice, Sturova 31, 080 01 Presov, Slovakia; patrik.fejko@tuke.sk (P.F.); robert.balint.bali@tuke.sk (R.B.B.)
[2] Faculty of Management and Business, University of Presov, Namestie Legionarov 3, 080 01 Presov, Slovakia; miroslav.gombar@unipo.sk
* Correspondence: jozef.dobransky@tuke.sk; Tel.: +421-55-602-6350

Abstract: Cataphoresis varnishing enables an organic coating to form on an aluminum substrate, thus increasing its corrosion resistance and durability. Cataphoresis varnishing is known to ensure a high adhesion of the created cataphoresis layer and a good homogeneity of this layer, even on surfaces with complex geometry. This paper aimed to optimize the deposition process and to analyze and evaluate the thickness of a cataphoresis layer formed on an aluminum substrate from AW 1050—H24 material. In total, 30 separate samples were created in accordance with the Design of Experiments methodology, using a central composite plan. The independent input factors in the study were: the electrical voltage (U) and deposition time in the cataphoresis varnishing process (t_{KTL}) at the polymerization times of 15 min, 20 min, and 25 min, respectively. The results of the statistical analysis showed that the voltage accounted for 33.82% of the change in the thickness of the created layer and the deposition time contributed 28.67% to thi change. At the same time, the interaction of the voltage and deposition time ($p < 0.0001$) accounted for 20.25% of the change in the thickness of the layer under formation. The regression model that was constructed showed a high degree of prediction accuracy (85.8775%) and its use as a function for nonlinear optimization provided a maximum layer thickness t_h of max = 26.114 μm, at U = 240 V and t_{KTL} = 6.0 min, as was proven under experimental conditions.

Keywords: cataphoresis; electrophoresis; coating layer thickness; analysis; planning conditions

1. Introduction

Electrophoretic paints, commonly known as electrocoats or paints, are organic coatings dispersed in water that carry an electric charge. This enables the paint to be used for deposition onto a metal that is carrying an opposite charge. Special needs for formulating this coating result from this special way of application [1–3].

Automotive coatings and the processes used to paint automotive surfaces exemplify avant-garde technologies capable of producing durable surfaces that exceed customer expectations for appearance, maximizing efficiency, and meeting environmental regulations. These achievements are rooted in 100 years of experience, trial and error techniques, technological advances, and theoretical evaluation [4].

The overall critical performance factors that drive the development and use of advanced automotive coatings and coatings technology are aesthetic properties, corrosion protection, mass production, cost, environmental requirements, appearance, and durability [5].

The adhesion of the coating to the material is also a very important factor for the appearance and durability of the surface of the material. Based on research, a torsional delamination test was developed, which consisted of applying an increasing torque on a hexagonal base directly glued to the coating. The test was quantitative and made it possible to calculate the shear stress that arose during delamination. Based on this test, it was found that polymerization temperature is an important factor in the adhesion of the material [6].

Anti-corrosion protection is also provided to ensure the durability of the coating. One of the technologies used to ensure this anti-corrosion protection is cataphoresis, which is used to apply paint to the paint surface. It is known for ensuring a high coating adhesion and good homogeneity, even for surfaces with complex geometry [7–9].

The researchers Mr. Rossi, Calovi, and Fedel conducted research for the implementation and optimization of the deposition process and the evaluation of the properties of a cataphoretic coating applied to an aluminum foam. They found a large effect for the corrosion behavior of the painted foam, which was evaluated using acetic acid salt chamber exposure and electrochemical impedance spectroscopy. By inserting the dye into the resin, it was possible to observe three types of cells, namely, black-colored cells that represented the coating; light-colored cells without traces of the coating resin; and cells with a purple color, which represented traces of resin. It was found that it is very difficult to obtain a uniform coating on the entire surface of aluminum according to the foam sample; another important factor is the deposition voltage, which achieves coatings with a greater thickness [10].

In further research, it was also confirmed that the higher the coating voltage, the greater the thickness of the layer. However, the cataphoresis process appears to be a promising technique for coating a material surface. If we set a relatively smaller coating voltage, it is possible to obtain a relatively thin coating, while it is necessary to avoid exceeding the coating voltage to values that are too high, which can lead to the formation of bubbles [11–13].

Other methods of applying organic coatings include adding graphene oxide to a cataphoresis bath. Research has found that graphene oxide leads to the formation of defective layers, with the consequence of reducing the durability of the coating. However, when applied in two steps with two different baths, it is possible to maintain the integrity of the coating and ensure the protection of the substrate. In the first bath, an epoxy-based method has been used, where an epoxy resin was used, which ensured an excellent level of material adhesion and good mechanical properties of the coating. The second bath contained graphene oxide, also called the black bath. From this research, we can determine only one thing: that the black bath guarantees a much greater thickness of the layer, thereby guaranteeing excellent protection [14–17].

In further research, the authors looked at the compatibility between the cataphoretic electrocoating and a silane surface layer. The research was carried out on a sheet of steel that had previously been treated with a silane sol-gel. In the case of thin samples coated with 120 nm silane sol-gels, the electrodeposition conditions were slightly affected. On the contrary, at a greater thickness, degradation occurred due to hydrogen production and bubbling [18–20].

The authors see the present paper as a contribution to the procedural approach to the complex process of creating anti-corrosion layers, such as the process of cataphoresis varnishing. Since the technological processes of surface treatment represent multifactor systems with interacting physical, chemical, and technological effects and, at the same time, since their influences may be considered random variables, the authors subjected the experimentally obtained data to proper methods of statistical analysis to gain a deeper understanding of these interrelationships. Another undisputed benefit of the present paper is the nonlinear optimization (maximization) of the basic technological parameter, the thickness of the created layer, and the analysis of the rate of deposition in the process of cataphoresis varnishing. However, the limitations are those of the experiment constraints and the use of only two basic process factors.

2. Materials and Methods

With the need to minimize costs and time and, at the same time, to maximize the reliability and objectivity of the information obtained about the process of anodic aluminum oxidation, it was necessary to conduct the experiment with as few trials as possible. The experimental planning methodology—DoE (Design of Experiments)—was used for the experimental verification of the influence of the basic process factors on the thickness of the layer created by the cataphoresis. This methodology represented the only scientifically justifiable methodology of experimentation and allowed for an obtainment of the maximum amount of information with a high statistical and numerical correctness, i.e., with a high reliability of the implemented conclusions and an optimal (minimum) number of individual trials. For the purpose of the experimental verification of the influence of the selected process factors (deposition time, varnishing voltage, and polymerization time), a Central Composite Plan was chosen, which facilitated the creation of a non-linear model, which we assume, in view of our practical experience, to be the case. The total number of trials, in terms of the type of the plan used, was 10. Since the marginal intention was also to examine the influence of the polymerization time (t_{pol} = 15 min, 20 min, and 25 min) at a constant polymerization temperature (T_{pol} = 200 °C) on the thickness and adhesion of the cataphoresis varnish, the experiment was carried out in three separate blocks. Each block represented one polymerization time [21,22].

The basic knowledge of the technological process and the method of its management could be obtained through the method of a factor experiment. The result of the factor experiment was an interpolation model that had the form of a first- or higher-degree polynomial (linear or non-linear model). In the analysis, we obtained results that allowed us to discover stages of the technological process. The planning of the technological process for the linear model can be written with the mathematical equation:

$$\hat{y} = b_0 + \sum_{j=1}^{k} b_j \cdot x_j \qquad (1)$$

Surface areas with higher-order models can be described more accurately if it is not possible to create an adequate linear model. We can write the technological process planning for the non-linear model using the mathematical equation:

$$\hat{y} = b_0 + \sum_{j=1}^{k} b_j \cdot x_j + \sum_{j \neq g=1}^{k} b_{jg} \cdot x_j \cdot x_g + \sum_{j=1}^{k} b_{jj} \cdot x_{jj}^2 \qquad (2)$$

Within relations (1) and (2), $\hat{y}$ represents the estimate of the investigated parameter (the thickness of the created layer), x_i represents the independent variables (U_{KTL}, t_{KTL}, and t_{pol}), and b_0, b_i, b_j, b_{jg}, and b_{jj} represent the estimates of the regression coefficients, which were calculated based on the method of least squares.

To reduce the number of experiments and utilize the results of the linear model experiment, we used composition plans. According to the location of the points, we divided composition plans into central and non-central [23].

2.1. Material Selection

In the elaboration of the work, namely, the analysis of the effect of the cataphoresis coating process conditions on the layer quality of the aluminum parts, aluminum specimens of the AW–1050 H24 type were used. Today, aluminum is used to produce various types of components for the automotive industry. EN AW–1050 A is a non-alloy aluminum with a maximum impurity content of 0.5%. The material is thermally uncurable. Increasing its strength is possible only under cold conditions (by rolling and pulling, etc.), when the increase in this strength is related to a reduction in elasticity and, thus, formability. In the soft annealed state (0), the material has an excellent formability (by bending and deep drawing, etc.). In the hardened states of H14 and H24, this formability is substantially lower. It should be taken into account that the H24 state exhibits a slightly better formability than that of the H14 state. It is used, among others, in the production of storage tanks, heat exchangers, spotlights, and packaging materials, etc. Its corrosion resistance is excellent under normal atmospheric conditions. This can be improved by the products' technical anodic oxidation. Non-alloy aluminum is very weldable using all common aluminum welding procedures (especially the MIG and TIG gas arc welding procedures). Under certain conditions, it may be necessary to soft anneal the material. This soft annealing temperature is from 320 to 350 °C [24–26].

2.2. Technological Process of Production

The experiment was carried out as part of a production operation on an automated cataphoresis line. The preparation of each sample, before the actual cataphoresis coating, was as follows:

(a) Chemical degreasing of the samples—chemical degreasing was carried out in a high alkaline medium emulsifying agent containing low-foaming tensides. Each sample was subjected to a degreasing time of 8 min under a constant temperature of 65 °C with a constant chemical composition of the solution (40 g·L^{-1}). The degreasing was followed by a two-stage rinse in demineralized water.
(b) Pre-phosphating activation—this was carried out in a commercial preparation from Pragochema CZ, trade name Pragofos 1927, under the following constant conditions: pH = 9.5, T_{act} = 40 °C, and t_{act} = 2 min.
(c) Phosphating—the phosphating itself was carried out in a multi-cation phosphatizing solution without nitrite accelerators. The samples were phosphated under constant operating conditions: T_{ph} = 50 °C and t_{ph} = 5 min and a constant chemical composition: a total Fisher spot content of 16 points, a free acid content of 0.8 g·L^{-1}, an accelerator content of 2.3 g·L^{-1}, a zinc content of 0.95 g·L^{-1}, and a phosphate content of 12.5 g·L^{-1}, with a pH of 3.40. The surface weight of the deposited phosphate coating ranged from 1.97 to 2.09 g·m^{-2}. The surface homogeneity after phosphating can be seen in the image of the checking sample shown in Figure 1. The SEM images were captured using a Scanning Electron Microscope Tescan Mira 3 FE equipped with an integrated EDX analyzer from Oxford Instruments, which allowed for an observation of the microstructure of the material and the performance of an elemental analysis (spot and surface distribution). For the SEM images, the secondary electron mode (SE) and an accelerating voltage of 15 kV were used. The distance between the sample and detector was 15 mm and the view field was 185 µm. A three-stage rinse in demineralized water followed the phosphating process.
(d) Cataphoresis varnishing—this was carried out according to the matrix of the experiment plan using a central composite plan. The basic variable factors are presented in Table 1. In the cataphoresis varnishing of the individual samples, the constant temperature of the cataphoresis paint was T_{KTL} = 32.5 °C and the value of the current flowing through the electrochemical system, namely I_{KTL} = 200 A, was kept constant. The chemical parameters of the cataphoresis paint during the experiment were also maintained at a constant level: dry matter (1 h at 110 °C) at 15.300, P/B ratio (binder/paint) at 0.151, pH (at 25 °C) at 5.82, and conductivity (at 25 °C) at 1660 µS·cm^{-1}.

Table 1. Values of variable input factors.

Factor Code	Factor	Unit	Factor Level				
			−2	−1	0	+1	+2
x_1	U_{KTL}	V	200	220	240	260	280
x_2	t_{KTL}	min	3.0	4.5	6.0	7.5	9.0
x_3	t_{pol}	min		15	20	25	

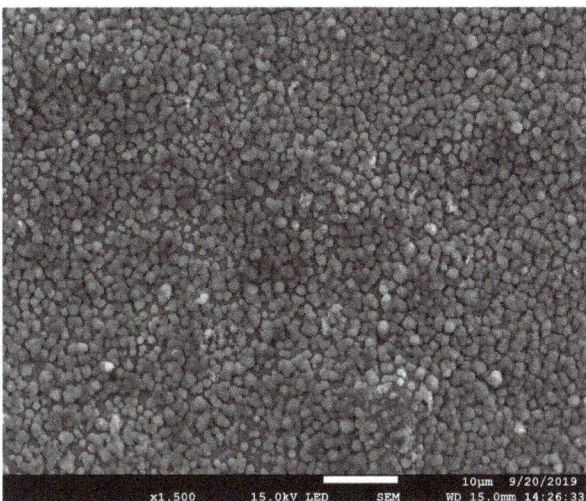

Figure 1. Homogeneity of the phosphate layer during the experiment (m_S = 1.97 g·m^{-2}).

In the actual implementation of the experiment, the Design of Experiments methodology was used, including the selection of the central composite plan. Table 1 shows the basic levels of the experiment plan for the individual input factors where, through their combination, individual experiments were carried out. The particular levels of the input factors were selected based on the practical experience of the authors.

2.3. Thickness Measurement

The layer thicknesses on the individual aluminum specimens were measured with the Elcometer 456 digital thickness gauge. This apparatus autonomously evaluated the average value of the coating when measured at specified points. The coating thickness ranged from 15 μm to 70 μm on the individual specimens. The measurement itself produced three types of measurement errors, i.e., systematic measurement errors, which were detected from the statistics, random errors (could not be influenced, they occurred during the measurement and were caused either by a failure to clean the surface of the components or by the influence of the temperature fluctuations, etc.), and gross errors (caused by observer fatigue and inattention) [27].

3. Results and Discussion

The process of cataphoresis varnishing can be viewed from two angles. The first is the electro-osmotic theory of cataphoresis. It is assumed that an electric bi-layer emerges at the interface between the solid and liquid phase. A part of this double layer is deposited as a liquid coating on top of the solid phase and the other part is scattered in the adherent liquid layer. As long as the solid phase can move freely in the liquid, the tangential component of the electrical force sets the suspended particles into motion. Cataphoresis varnishing uses the principle of cathodic organic coating creation based on epoxy or acrylic cataphoresis materials. Water-soluble cationic coatings with very low organic solvent content contain particles of varnish in the form of polymer cations. Thus, if an electric field is created in this system, with the solid phase particles scattered in the liquid phase, the particles begin to move in the direction of the electric field under the influence of the electric force. A direct current between the coated part, which is the cathode in the cataphoresis varnishing process, and an anodic counter electrode (anode) creates an electric field that becomes the carrier of the polymer cations that travel towards the cathode. In the course of the reactions with hydroxyl ions resulting from the breakdown of water, the solubility is suppressed on the cathode, and the organic coating deposition process is activated on the surface of the cathode. The second view of the cataphoresis varnishing process is the ionic theory of cataphoresis. In this theory, suspended particles are considered to be high-molecular-weight electrolyte molecules. These molecules then disassociate into high-power ions and associated electrolytic ions, which carry the same amount of electrical charge, but of the opposite polarity. The moving particles of the solid phase, which are suspended in the liquid phase under the influence of the electric field, are seen as electrolytic ions in electrolyte solutions. The electric charges of the ions are affected by an electric force in the electric field, the magnitude of which is determined by the product of the magnitude of the electric charge and the magnitude of the electric field. This force accelerates the ion, which is, at the same time, hampered by the movement of the frictional force emerging in the liquid environment. The ion is, at this time, considered to be a sphere with a radius corresponding to its resistance in the given environment, defined by the Stokes equation of resistance of a sphere in a liquid. The friction force is proportional to the velocity of the ion. Upon the introduction of the electrical charge, a steady state occurs. The mobility of the ions is directly proportional to the power and inversely proportional to the radius of the ion [28,29].

Since the implemented methodology of the experimental verification (Design of Experiments) represented a statistical approach, the subsequent analysis of the experimentally obtained data was, too, carried out using mathematical–statistical procedures. The initial analysis of the applied model pointed to the fact that the proportion of the variability of the measured thickness of the cataphoresis coating was 86.70225% and the adjusted index of determination, determining the degree of explanation of the data variability by the model, was 85.8775%. The average thickness of the cataphoresis layer formed, covering all the individual trial runs, was t_h = 24.464 ± 3.677 µm.

The table of the variance analysis (Table 2), as a basic requirement for the correctness of the regression model, enabled us to conclude that the variability caused by random errors was significantly lower than the variability of the measured values explained by the model, and the value of the achieved significance level (p) indicated the adequacy of the model used, based on the Fisher–Snedecor test criterion. Another view of this analysis is through assessing the adequacy of the model itself and is based on the very essence of the variance analysis. For testing the null (H_0) statistical hypothesis, which followed from the nature of the test and said that none of the effects (factors) used in the model effected a significant change on the examined variable, it followed that the achieved level of significance (p) was less than the selected level of significance α = 0.05, and it could be concluded that we did not have enough evidence to accept H_0 and we could say that the model was significant [30].

Table 2. The table of variance analysis.

Source	df	SS	MS	F	p
Model	5	3956.890	791.378	58.5323	<0.0001 *
Error	144	1946.932	13.52		
C. Total	149	5903.821			

SS—Sum of Squares, MS—Mean squares, F—Fisher's test statistic, p—achieved level of significance, and *—significant at the level of significance α = 0.05.

The applied model was further tested in the so-called insufficient model adaptation error test (Table 3), where we tested the scatter of the residues and scatter of the data measured within the groups; thus, we tested the premise of whether the regression model adequately described the observed dependence. Based on the error test of insufficient model adaptation, due to the achieved significance level of 0.1853, a zero statistical hypothesis could be accepted at the selected significance level of α = 5% and it could be said that the scatter of the residues was less than or equal to the scatter within the groups and, therefore, the model could be considered sufficient.

Table 3. Model fit error.

Source	df	SS	MS	F	p
Lack Of Fit	3	283.9322	94.6441	8.0245	0.1853
Pure Error	141	1663.000	11.7943		
Total Error	144	1946.932			

SS—Sum of squares, MS—Mean square, F—Fisher's test statistic, and p—achieved level of significance.

Based on the above assumptions and their fulfillment (Tables 2 and 3), the following table (Table 4) presents an estimate of the regression model parameters with testing the significance of the individual effects and their combination at the significance level $\alpha = 0.05$.

Table 4. Estimates of regression model coefficients.

Term	Estimate	Std Error	t	p	−95% CI	+95% CI
Intercept	26.114	0.567	46.030	<0.0001 *	24.993	27.236
x_1	3.170	0.274	11.560	<0.0001 *	2.628	3.711
x_2	2.686	0.274	9.800	<0.0001 *	2.144	3.227
$x_1 \cdot x_2$	3.286	0.475	6.920	<0.0001 *	2.347	4.224
$x_1 \cdot x_1$	−0.474	0.233	−2.030	0.0439 *	−0.934	−0.013
$x_2 \cdot x_2$	−0.902	0.233	−3.870	0.0002 *	−1.362	−0.441

x_1—voltage (V), x_2—deposition time (min), t—Student's test criterion, p—achieved level of significance, CI—confidence interval, and *—significant at the level of significance $\alpha = 0.05$.

The results shown in Table 4 thus enabled the building of a predictive mathematical-statistical model at a coded scale:

$$y(t_h) = 26.114 + 3.170 \cdot x_1 + 2.686 \cdot x_2 + 3.286 \cdot x_1 \cdot x_2 - 0.474 \cdot x_1^2 - 0.902 \cdot x_2^2 \quad (3)$$

Since the DoE methodology worked with a code scale, in order to ensure the numerical and statistical correctness of the results, it was necessary to convert Equation (3) to the scale of the original variables, the natural scale. Considering that the code scale represented the DoE standardization of the variable input factors, it was necessary to use the following equation to convert to the natural scale:

$$X_d(i) = \frac{x(i) - \frac{x_{max} + x_{min}}{2}}{\frac{x_{max} - x_{min}}{2}} \quad (4)$$

where $x(i)$ represents the original basic variable, $i = 1, 2, \ldots, n$ is the number of basic factors, x_{max} is the maximum value of the original variable $x(i)$, and x_{min} is the minimum value of the original variable $x(i)$.

Thus, when using the regression model (3) in the code scale, taking into account the conversion Equation (4) for the individual variable input factors and subsequent adjustment, it was possible to make a notation of a prediction equation for the thickness of the cataphoresis layer in the form of:

$$th = 52.370 + 7.01 \cdot 10^{-2} \cdot U - 19.687 \cdot t_{KTL} + 0.109 \cdot U \cdot t_{KTL} - 1.185 \cdot 10^{-3} \cdot U^2 - 0.409 \cdot t_{KTL}^2 \quad (5)$$

The analysis of Table 4 showed that the largest share in explaining the variability in the parameter under study, the thickness of the cataphoresis layer per absolute element of the model (intercept), which was involved in changing the thickness value of the layer, was that of 57.387%. From a methodological point of view, the absolute element of the model was characterized by "neglected" influences, which we kept at a constant level in the experiment (especially the chemical characteristics of the cataphoresis electrolyte, current density, and anode-to-cathode ratio), or we did not consider them. If we neglected this model element and subsequently analyzed only the basic variable input factors, we would have come to the conclusion (Table 4) that the most significant factor that affected the thickness of the layer formed was the voltage (*U*). It accounted for 33.82% of the change in the thickness of the layer. The second most significant element of the model (5) was the deposition time in the cataphoresis varnishing process (t_{KTL}), accounting for 28.67% of the change in the thickness. At the same time, the interaction of the voltage and deposition time accounted for 20.25% of the change in the thickness of the layer formed. The nonlinear model elements (5), namely, the voltage squared and the deposition time squared, accounted for 5.94% and 11.32%, respectively, of the change in the thickness of the created layer. As seen in Table 4 and model (5), it was clear that the processes of the surface treatment of the metals, including the cataphoresis varnishing, were best described by non-linear models with a significant influence and mutual interaction of the individual factors. The model (5) also needed to be expanded and modified by the influence of the chemical factors acting in the process of the cataphoresis varnishing. The model (5) represented a steppingstone to a comprehensive analysis of the cataphoresis varnishing process using a statistical approach. The statistical approach was chosen because the studied layer parameters were understood as random variables in the mathematical sense [31].

The plotted thickness of the cataphoresis layer formed during the respective deposition times in the cataphoresis coating process under various voltages is shown in Figure 2.

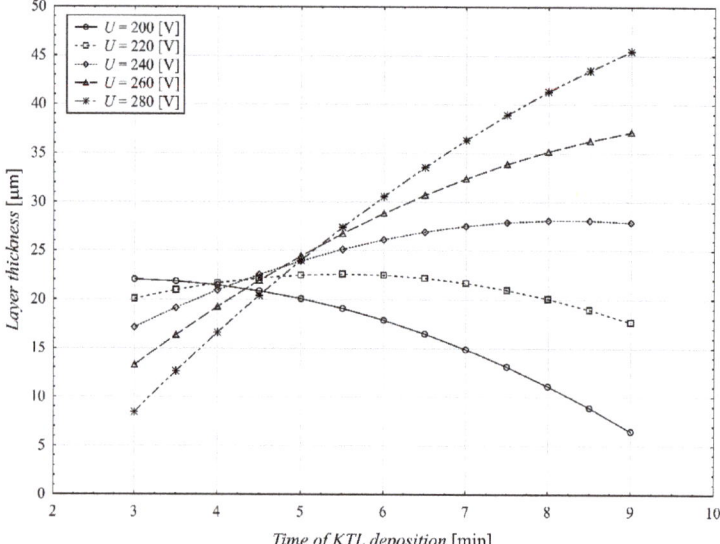

Figure 2. Dependence of the thickness of the formed layer on the time of KTL deposition at different voltage values.

The graph shows that, by increasing the deposition time of the varnishing medium, the thickness of the layer formed was reduced under different varnishing voltages. A varnishing voltage of 200 V and a deposition time of the varnishing medium of 3 min affected the layer thickness the most. The thickness of the layer increased during the 3 min deposition period, after which, the thickness of the layer decreased. This was due to a low electric current, which caused the coagulation of the paint on the surface where it stopped; thus, the coated part became electrically non-conductive. With an increased varnishing medium deposition time, the value of the electric current decreased due to an increase in the thickness of the deposited layer t_h, and the value of the current decreased to zero.

Under a 220 V voltage, the thickness of the coating increased for 5.5 min when the coating also reached its maximum thickness. After this time, the thickness of the coating decreased. Increasing the varnishing voltage to 240 V meant increasing the deposition time of the varnishing medium up to 8 min, without significantly affecting the thickness of the layer. A further increase in the varnishing voltage resulted in an accelerated layer formation and a similar change was observed when the varnishing voltage was increased to 260 V and 280 V, when the thickness of the layer reached its maximum values throughout the deposition time of the varnishing medium t in the electrolyte. This phenomenon could be attributed to the color deposition technology that ran in the following sequence: water electrolysis, ion migration (electrophoresis), the coagulation of the polymer on the cathode, and water ejection via osmotic pressure. This phenomenon, as such, could be explained by Ohm's law, which says an electric current of a constant voltage is created between a cathode and anode, which can be described by the equation U = R*I. Cataphoresis varnishing works on the principle of creating cathodic organic coatings based on epoxy materials. Cationic coatings soluble in water contain a small number of organic solvents and, at the same time, particles in the form of polymer cations.

Once the coating was deposited on the cathode in the process of cataphoresis, the resistance reached its maximum values. The layer ceased to be conductive and became insoluble in water again. This deposited layer needed to be subsequently cured in a reaction with another polymer. At this stage, hydroxyl groups along the molecular chain in the cationic resin were applied, which reacted with isocyanates (they were equally present in the resin) to form urethane compounds.

The function gradient (5), i.e., the direction of the steepest addition to the layer thickness under the input parameters examined, namely the voltage (U) and the deposition time in the course of the cataphoresis varnishing (t_{KTL}), is defined by the vector:

$$\nabla t_h(U, t_{KTL}) = [0.10953 \cdot U - 0.80176 \cdot t_{KTL} - 19.6867; 0.10953 \cdot U - 0.0037 \cdot t_{KTL} + 0.0701] \tag{6}$$

The relationship (6) thus defines the direction, depending on the input variables, in which the function (5) grew the fastest, that is, the direction where the thickness of the forming layer reached its maximum in the shortest possible time.

In terms of the cataphoresis layer formation, based on the mathematical–statistical models (5) and (6), it is possible to define the layer formation rate as the first derivative of the function (5), according to the deposition time (t_{KTL}):

$$v_{th} = \frac{\partial t_h}{\partial t_{KTL}} = 0.10953 \cdot U - 0.80176 \cdot t_{KTL} - 19.6867 \tag{7}$$

Thus, Equation (7) represents the direction of the steepest rise in the function (6) in the direction of the voltage. Thus, from Equation (7), it follows that the rate of formation of the cataphoresis layer under the given experimental verification conditions (Table 1) was a function of the voltage and deposition time in the process of the cataphoresis varnishing. In accordance with theoretical knowledge and on the basis of Equation (7), we can conclude that, by increasing the voltage, the rate of deposition of the cataphoresis layer also increased, and, on the other hand, by increasing the deposition time, this rate in the cataphoresis varnishing process decreased. The decrease in the rate of the formation of the cataphoresis layer and the influence of the deposition time depended on the electrical properties of the forming layer. Considering the fact that the layer formed during cataphoresis was electrically non-conductive, its electrical resistance must have inevitably increased with an increase in its thickness; therefore, the rate of its formation must have decreased. However, if we wanted to ensure a constant rate of formation of the layer throughout the entire deposition period in the cataphoresis varnishing process, we would have to increase the voltage in proportion, as per Equation (7). The rate of the formation of the layer is a fairly important indicator of the cataphoresis varnishing process. However, there are two opposing requirements of the rate of the layer formation. On the one hand, there is a requirement to achieve the highest possible rate of layer formation, thereby reducing the time required for the cataphoresis varnishing to run its course, which results in an increased economic efficiency of the process itself. The counter requirement stems from the process of layer formation in relation to its quality. If the rate of the layer formation is too high, the hydrogen that emerges on the surface of the treated structural part does not have enough time to "escape" the surface, and the resulting layer "traps" it in the surface of the part. However, in the process of polymerization, this trapped hydrogen creates defects in the layer in the form of craters. Further research is needed to determine the optimal value for the rate of deposition, taking into account the basic requirements above [32].

Equation (7) is a statistical equation and, therefore, within, it holds only the input variables' intervals and the factors used (Table 1). Its extrapolation beyond these factor values intervals may lead to incorrect results and conclusions.

The second partial part of the analysis, shown in Table 1, was devoted to the evaluation of the thickness of the cataphoresis layer in relation to the polymerization time (t_{pol}), using three different times as part of the experiment plan, namely 15 min, 20 min, and 25 min, respectively, for each combination. The basic graphical representation of the influence of the polymerization time on the thickness of the layer created in individual combinations of the input factors (U, t_{KTL}) is shown in Figure 3.

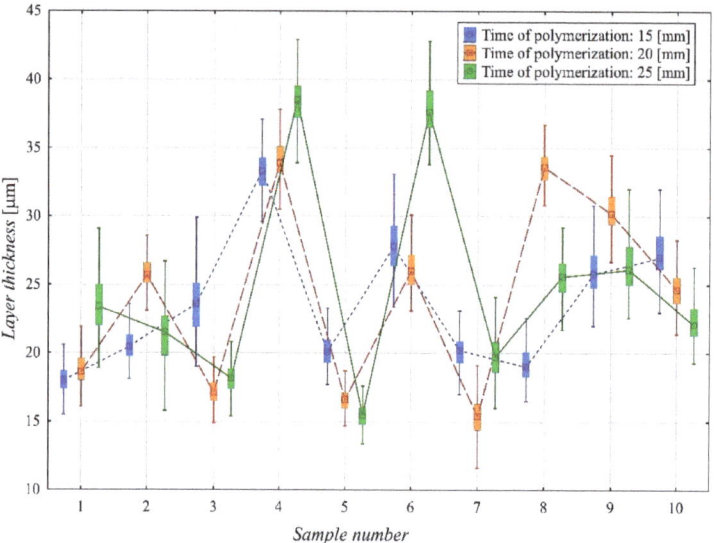

Figure 3. Effect of polymerization time on the thickness of the cataphoretic layer for individual experiments.

Figure 3 makes it evident that the polymerization time affected the resulting thickness of the cataphoresis layer in a relatively random manner. However, the polymerization process itself showed that, depending on the type of cataphoresis paint used, 10% and 20% of it was lost in the polymerization process. This was because the polymerization process did not directly participate in the formation of the cataphoresis layer, but affected its resulting properties. The average thickness of the cataphoresis layer formed after the polymerization at a constant temperature of 200 °C and a polymerization time of 15 min was 23.709 ± 0.192 µm. Here, it is necessary to say that in this analysis, all the values of the measured thickness were used, including repetitions of individual measurements (7565 measurements). For a polymerization time of 20 min at a constant temperature of 200 °C, the average thickness value of the formed layer was 24.349 ± 0.267 µm, and for a polymerization time of 25 min, the average thickness of the layer was 24.937 ± 0.289 µm.

Thus, the average difference in the thickness of the cataphoresis layer between the individual polymerization times was 0.613 ± 0.103 µm between the polymerization times of 20 min and 15 min, 0.619 ± 0.102 µm between the polymerization times of 25 min and 20 min, and finally, 1.220 ± 0.102 µm between the polymerization times of 25 min and 15 min (Figure 4).

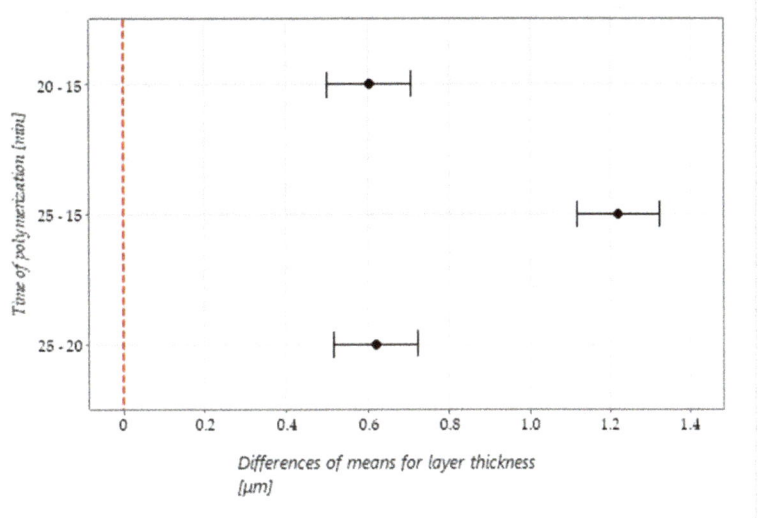

Figure 4. Values of differences in the thicknesses of the layer created by cataphoretic painting for uniform polymerization temperatures.

A graphic representation of the model verification (5) under the practical conditions of the production process is shown in Figure 5. The verification was carried out at U_{KTL} = 240 V on the same samples, listed in the Material Selection and Technological process of production section, and under the same conditions as the main experiment.

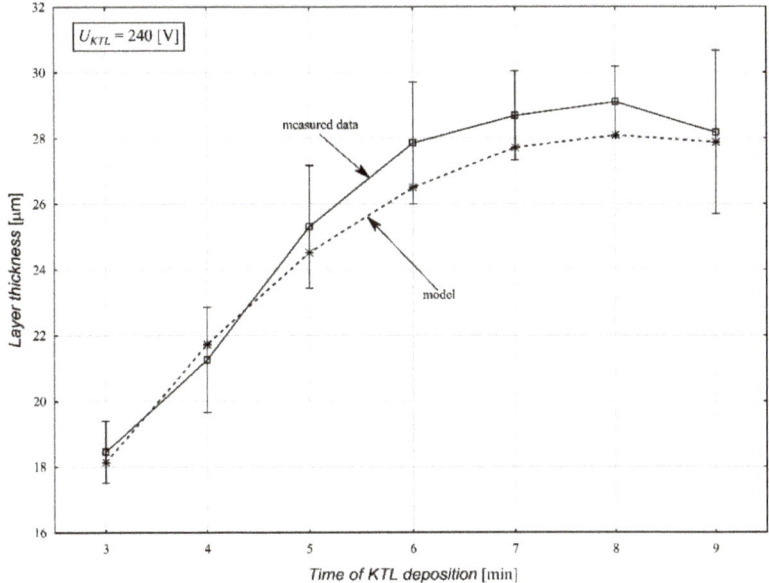

Figure 5. Graphical comparison of the thickness of the created layer within the verification experiment and the thickness of the layer calculated by the prediction model (5).

As part of the analysis of the modelled thickness values of the cataphoresis layer created and the values obtained from repeated measurements of the verification experiment, we came to conclusion that the average deviation in all the measurements carried out was 0.632 μm (2.14%), while the lowest negative deviation in the calculated thickness of the layer measured and the lowest deviation in the model (5) was at the level of −1.801 μm (8.690%), and the maximum positive value of the examined difference was at the level of +2.306 μm (8.407%). At the same time, based on the Shapiro–Wilks test, it can be said that the residues showed a normal Gaussian distribution (p = 0.233) at the selected level of significance, which indicated that the model (5) also met the last condition for the regression triplet analysis and could be considered correct. A graphical representation of the differences between the measured and modeled values of the thickness of the created layer is shown in Figure 6.

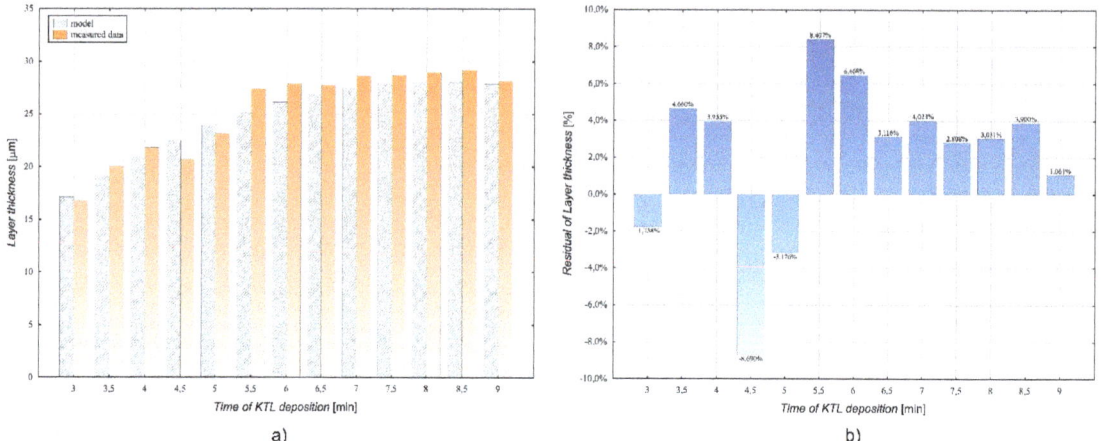

Figure 6. Differences between the measured and calculated thickness of the cataphoretic layer ((**a**) comparison of the results of the verification experiment and the model (5), and (**b**) percentage display of the residues for the verification experiment).

The morphology of the surface was also of significant importance from the point of view of the quality of the created cataphoretic layer. The change in the morphology and structure of the surface of the created layer depended primarily on the conditions of the process of creating the layer, that is, on the operation of the cataphoretic painting itself.

Figure 7 shows the surfaces of the layers created at voltages of U_{KTL} = 220 V, 260 V, and 280 V with deposition times of t_{KTL} = 7.5 min and 6.0 min at a constant polymerization temperature of 200 °C, but with a different polymerization times: 15 min (a), 20 min (b), and 25 min (c). It is clear from the mentioned morphologies that the tension in the process of creating the cataphoretic layer had a significant influence. At a voltage of 220 V, in all cases (a, b, c), a relatively smooth surface without a distinct structure was scanned. At a voltage of 260 V, morphological changes began to appear in the form of a slightly distinct structuring of the surface, but in the presence of a significant defect in the form of craters. These craters could be attributed to the process of cataphoretic painting in the form of the binding of the hydrogen on the surface of the painted sample with its binding by the created layer and subsequent "explosion" in the polymerization process. However, this defect was no longer observed when using a voltage of 280 V, but the surface of the created layer already had a pronounced wrinkled structure. In general, it can therefore be said that, by increasing the tension in the process of creating a layer, the morphology of the created layer deteriorated and the created layer acquired a significantly wrinkled structure.

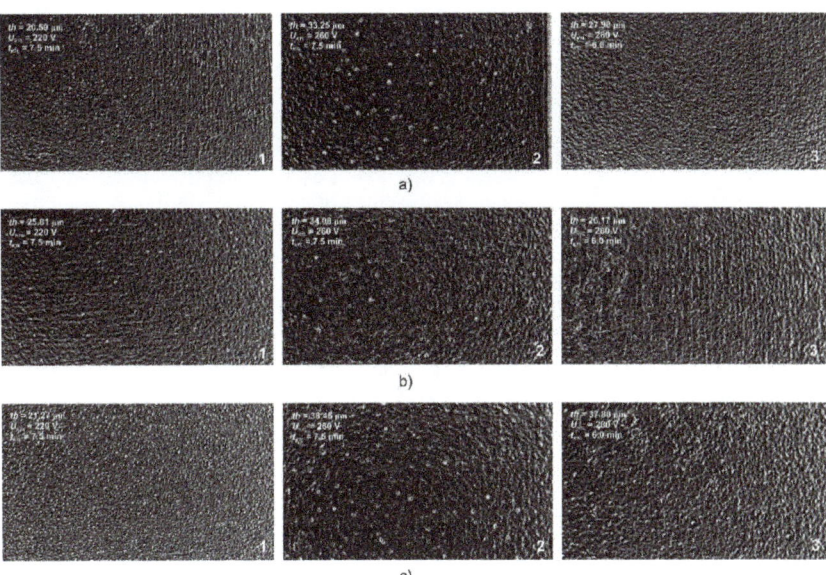

Figure 7. Morphological changes in the surface of the created cataphoretic layer depending on the voltage and the deposition time ((**a**) t_{POL} = 15 min, (**b**) t_{POL} = 20 min, and (**c**) t_{POL} = 25 min; 1—U_{KTL} = 220 V, t_{KTL} = 7.5 min, 2—U_{KTL} = 260 V, t_{KTL} = 7.5 min, and 3—U_{KTL} = 280 V, t_{KTL} = 6.0 min).

The confirmation of the above conclusions was carried out using additional experiments at a voltage of 400 V and deposition times of 5.0 min and 4.0 min, while the morphology of the surface of the cataphoretic layer is shown in Figure 8. It is obvious that the surface morphology of the formed layer at a high voltage was significantly structured with very pronounced wrinkling; however, with a deposition time of 5.0 min, the surface was significantly more heterogeneous than that in the case of a deposition time of 4.0 min. Thus, in addition to the applied voltage, the deposition time of the KTL process also had an effect on the surface morphology of the created cataphoretic layer and, with an increase in the deposition time, a more pronounced heterogeneity of the surface occurred.

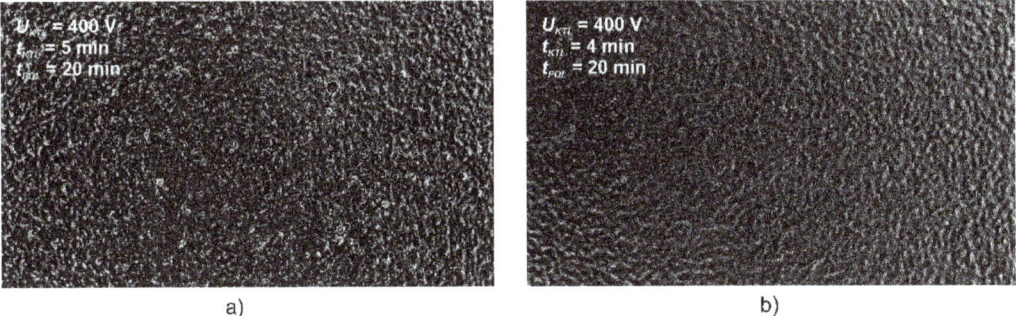

Figure 8. Morphological changes in the surface of the created cataphoresis layer at U_{KTL} = 400 V, t_{POL} = 20 min a T_{POL} = 200 °C ((**a**) t_{KTL} = 5 min, and (**b**) t_{KTL} = 4 min).

An important consequence of the defined predictive dependence (5) was a determination of the optimal values of the analyzed input variables (U, t_{KTL}). Due to the technological requirements placed on the thickness of the layer under formation, it was advisable to look for the maximum regression function (5). The general optimization problem was to select n decision variables $x_1; x_2; \ldots ; x_n$ from a given implemented area, in such a way as to optimize (minimize or maximize) the purpose function:

$$f(x_1, x_2, \ldots, x_n) \tag{8}$$

The optimization problem was a non-linear programming problem (NLP) if the purpose function was nonlinear or the implemented area was defined by nonlinear constraints. Then, the maximization of the general nonlinear programming is defined in the form of:

$$max f(x_1, x_2, \ldots, x_n) \tag{9}$$

for restrictions:

$$g_1(x_1, x_2, \ldots, x_n) \leq b_1$$
$$g_2(x_1, x_2, \ldots, x_n) \leq b_2$$
$$\ldots\ldots\ldots\ldots\ldots\ldots\ldots\ldots\ldots\ldots \tag{10}$$
$$g_m(x_1, x_2, \ldots, x_n) \leq b_m$$

where each of the constraints g_1 through g_m is defined. A special case is linear programming. The obvious relation for this case is:

$$f(x_1, \ldots\ldots, x_n) = \sum_{j=1}^{n} c_j \cdot x_j \tag{11}$$

and

$$g_i(x_1, \ldots, \ldots, x_n) = \sum_{j=1}^{n} a_{ij} \cdot x_j i = (1, 2, \ldots, m) \tag{12}$$

Non-negative variables constraints can be included simply by attaching additional constraints:

$$g_{m+i}(x_1, x_2, \ldots, x_n) = -x_i \leq 0 i = (1, 2, \ldots, n) \tag{13}$$

In some cases, these constraints are considered explicit, as is any other issue in the delimited areas. In other cases, it is appropriate to consider them implicit if the non-negative constraints are manipulated, as is the case with simplex methods.

To simplify the proposition, let x denote the vector of the control variables $x_1, x_2, \ldots, x_n$, which represents $x = (x_1, x_2, \ldots, x_n)$. The problem is more aptly written in the form:

$$max f(x) \tag{14}$$

according to the:

$$g_i(x) \leq b_i (i = 1, 2, \ldots, m) \tag{15}$$

As in solving the tasks of linear programming, there are no restrictions on these formulations. When maximizing the $f(x)$ functions and, of course, also when minimizing its $f(x)$, the conditions of equality $h(x) = b$ can be written as two separate conditions of inequality, $h(x) \leq b$ and $-h(x) \leq -b$.

To optimize the thickness of the created cataphoresis layer, a regression model (5) was applied as a functional function and the interior point method was used for the nonlinear optimization. The ranges of the intervals of the variable input factors used were the basic constraints (Table 1), which are defined as follows:

$$200 \leq U \leq 2803 \leq t_{KTL} \leq 9 \tag{16}$$

A MATLAB software product Optimization Toolbox was used to implement the optimization of the thickness of the cataphoresis layer created. The task of the nonlinear optimization, in our case, was to find the maximum of the problem, which is defined as:

$$max f(x) \begin{cases} c(x) \leq 0 \\ ceq(x) = 0 \\ A \cdot x < b \\ Aeq \cdot x = beq \\ lb \leq x \leq ub \end{cases} \quad (17)$$

where x, b, beq, lb, and ub are vectors, A and Aeq are matrices, $c(x)$ and $ceq(x)$ are vector functions, and $f(x)$ is a scalar function. The course of the optimization process itself, as an output from the optimization program, is shown in Figure 9.

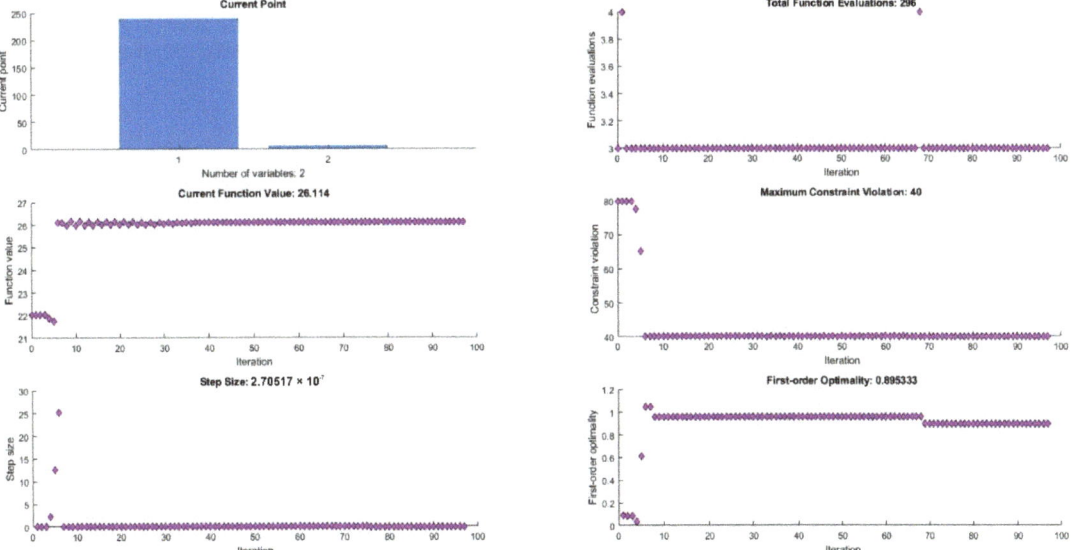

Figure 9. The course of optimization of the thickness of the created layer.

The result of the non-linear optimization process of the cataphoresis varnishing, considering only two variable technological factors, different voltages and deposition times in the process of the cataphoresis varnishing, was the determination of the maximum thickness of the purpose-built regression function (5). The maximum of the purpose-built function, while respecting the constraints given by Equation (16), was t_{hmax} = 26.114 µm under the following technological conditions: U = 240 V and t_{KTL} = 6.0 min. Therefore, in order to create the thickest layer possible, it was necessary to set these basic factors at a defined level [33].

However, we must also define the limitations of the conducted experimental research. The conclusions of the submitted study are valid only in the range of the experimental conditions listed in Table 1, which resulted from the applied statistical approach. A further limitation is imposed by the other relevant input conditions in the processes of degreasing, activation, and phosphating. Therefore, it will be necessary to expand the model (5) by including these impacts, thus defining the complex technological dependence of the process factors on the layer forming.

4. Conclusions

Today, cars are more than a means of transport for many, because they also create an image of the owner. Therefore, it is important what the vehicle looks like, which places demands not only on its design, but also on its surface treatment. Today's customers demand that its bodywork resists not only corrosion, which is achieved by using a good surface finish and high-quality varnishes, but also weather conditions (hail damage). Resistance to chemical influences that affect this bodywork, whether this is road salt or acid rain in winter, is also important. Progress in surface finishes and varnishing systems is constantly advancing. We can see this if we compare the technologies used 30 years ago to those used today. The treatment methods we use now are much more effective and environmentally friendly. This trend can also be observed in the surface treatment of bodywork, where, for example, the use of hexavalent chromium, which is toxic, is avoided. Great demands are placed on occupational safety, which is why quality and safe workplaces are essential. The varnishes and color shades used have also undergone a big transformation. The aim of the experimental part of this study was to create a planned experiment, on the basis of which, we analyzed the effects of varnishing voltage, varnishing current, deposition time, and layer thickness on the material surface. In total, 30 test samples of AW 1050—H24 were used. All the samples were passed through an automated cataphoresis varnishing line, from chemical degreasing to curing (polymerization). The thickness was then measured on the samples using a digital thickness gauge.

The initial analysis of the applied model pointed to the fact that the proportion of variability in the measured thickness of the cataphoresis coating was 86.70225% and the adjusted index of determination, determining the degree of explanation of data variability by the model, was 85.8775%. The average thickness of the layer formed in the process of cataphoresis, spanning all the individual trial runs, was t_h = 24.464 $\pm$ 3.677 μm. Based on the analysis of variance, it can be said that the variability caused by random errors was significantly less than the variability in the measured values. Based on the model of significance achieved, this indicated that this model is suitable for use. Under the voltage of 220 V, the thickness of the coating increased for 5.5 min, when the coating also reached its maximum thickness. After this time, the thickness of the coating decreased. Increasing the varnishing voltage to 240 V meant increasing the deposition time of the varnishing medium up to 8 min, without significantly affecting the thickness of the layer. A further increase in the varnishing voltage resulted in an accelerated layer formation, and a similar change was observed when the varnishing voltage was increased to 260 V and 280 V, when the thickness of the layer reached its maximum values throughout the deposition time of the varnishing medium t in the electrolyte.

The authors see the submitted paper as a contribution to the procedural approach to such complex processes of creating anti-corrosion layers, such as the process of cataphoresis varnishing. Since the surface treatment technological processes represent multifactor systems with interacting physical, chemical, and technological effects and, at the same time, since the influences may be considered as random variables, the authors subjected the experimentally obtained data to the correct methods of statistical analysis for a deeper understanding of these interrelationships. Another undisputed benefit of the present paper is the nonlinear optimization (maximization) of the basic technological parameter, the thickness of the created layer, and the analysis of the rate of the cataphoresis coating deposition. However, its limitations are those of the experiment constraints and the use of only two basic process factors.

However, it should be remembered that the process of creating a cataphoresis layer is a complex physical and chemical process, where the formation of bonds between the individual components of the cataphoresis paint, under the influence of an electric current, plays an essential role. Although mathematical models for the formation and growth of the cataphoresis layer [2] describe the causes of and, at the same time, the inter-relations between the factors involved in the growth of the cataphoresis layer, in relatively great detail, the authors' effort was to simplify the prediction of the thickness of the layer based on the practically used input factors in the KTL process, namely, the change in the deposition time and stress in the cataphoresis coating. Another fact is that the resulting property, the quality of the layer formed by cataphoresis, is not determined only by its thickness. The quality of the cataphoresis layer is also the result of its other properties, such as corrosion resistance, adhesion, bending, impact resistance, and hardness, as well as its aesthetic properties expressed by gloss and shade. All of these properties can be influenced within the complex KTL process, starting with pre-treatment and ending with polymerization. However, as part of the present paper, we focused only on the analysis of the thickness of the created cataphoresis layer as a basic parameter, which is prescribed in a customer's drawing documentation as a requirement for the painting process, which also affects the other, above-mentioned properties of the layer to some extent.

Author Contributions: Conceptualization, J.D. and M.G.; methodology, M.G.; validation, P.F.; formal analysis, R.B.B.; data curation, P.F. and R.B.B.; writing original draft preparation, J.D. All authors have read and agreed to the published version of the manuscript.

Funding: This research received no external funding.

Institutional Review Board Statement: Not applicable.

Informed Consent Statement: Not applicable.

Data Availability Statement: Not applicable.

Acknowledgments: This paper has been elaborated in the framework of the project KEGA 063TUKE-4/2021.

Conflicts of Interest: The authors declare no conflict of interest. The founding sponsors had no role in the design of the study; in the collection, analyses, or interpretation of the data; in the writing of the manuscript; or in the decision to publish the results.

References

1. Brüggemann, M.; Rach, A. *Electrocoat*; Vincentz Verlag: Hannover, Germany, 2020.
2. Goldschmidt, A.; Streitberger, H.J. *BASF-Handbuch: Lackiertechnik*; Vincentz Verlag: Hannover, Germany, 2002.
3. Brock, T.; Groteklaes, M.; Mischke, P. *European Coatings Handbook*, 2nd ed.; Vincentz Verlag: Hannover, Germany, 2010.
4. Akafuah, N.K.; Poozesh, S.; Salaimeh, A.; Patrick, G.; Lawler, K.; Saito, K. Evolution of the Automotive Body Coating Process-A Review. *Coatings* **2016**, *6*, 24. [CrossRef]
5. Skotnicki, W.; Jedrzejczyk, D. The comparative analysis of the coatings deposited on the automotive parts by the cataphoresis method. *Materials* **2021**, *14*, 6155. [CrossRef] [PubMed]
6. Mazeran, P.E.; Arvieu, M.F.; Bigerelle, M.; Delalande, S. Torsion delamination test, a new method to quantify the adhesion of coating: Application to car coatings. *Prog. Org. Coat.* **2017**, *110*, 134–139. [CrossRef]
7. Romano, A.P.; Olivier, M.G.; Vandermiers, C.; Poelman, M. Influence of the curing temperature of a cataphoretic coating on the development of filiform corrosion of aluminium. *Prog. Org. Coat.* **2006**, *57*, 400–407. [CrossRef]
8. Miskovic-Stankovic, V.B.; Stanic, M.R.; Drazic, D.M. Corrosion protection of aluminium by a cataphoretic epoxy coating. *Prog. Org. Coat.* **1999**, *36*, 53–63. [CrossRef]
9. Olivier, M.G.; Poelman, M.; Demuynck, M.; Petitjean, J.P. EIS evaluation of the filiform corrosion of aluminium coated by a cataphoretic paint. *Prog. Org. Coat.* **2005**, *52*, 263–270. [CrossRef]
10. Rossi, S.; Calovi, M.; Fedel, M. Corrosion protection of aluminum foams by cataphoretic deposition of organic coatings. *Prog. Org. Coat.* **2017**, *109*, 144–151. [CrossRef]
11. Poulain, V.; Petitjean, J.P.; Dumont, E.; Dugnoille, B. Pretreatments and filiform corrosion resistance of cataphoretic painted aluminium characterization by EIS and spectroscopic ellipsometry. *Electrochim. Acta* **1996**, *41*, 1223–1231. [CrossRef]
12. Almeida, E.; Alves, I.; Brites, C.; Fedrizzi, L. Cataphoretic and autophoretic automotive primers: A comparative study. *Prog. Org. Coat.* **2003**, *46*, 8–20. [CrossRef]
13. Deflorian, F.; Rossi, S.; Prosseda, S. Improvement of corrosion protection system for aluminium body bus used in public transportation. *Mater. Des.* **2006**, *27*, 758–769. [CrossRef]
14. Su, Y.; Zhitomirsky, I. Cataphoretic assembly of cationic dyes and deposition of carbon nanotube and graphene films. *J. Colloid Interface Sci.* **2013**, *399*, 46–53. [CrossRef] [PubMed]
15. Rudawska, A.; Wahab, M.A. The effect of cataphoretic and powder coatings on the strength and failure modes of EN AW-5754 aluminium alloy adhesive joints. *Int. J. Adhes. Adhes.* **2019**, *89*, 40–50. [CrossRef]
16. Calovi, M.; Dire, S.; Ceccato, R.; Deflorian, F.; Rossi, S. Corrosion protection properties of functionalised graphene–acrylate coatings produced via cataphoretic deposition. *Prog. Org. Coat.* **2019**, *136*, 105261. [CrossRef]

17. Rossi, S.; Calovi, M. Addition of graphene oxide plates in cataphoretic deposited organic coatings. *Prog. Org. Coat.* **2018**, *125*, 40–47. [CrossRef]
18. Zanella, C.; Fedel, M.; Deflorian, F. Correlation between electrophoretic clearcoats properties and electrochemical character-istics of noble substrates. *Prog. Org. Coat.* **2012**, *74*, 349–355. [CrossRef]
19. Fedel, M.; Druart, M.E.; Olivier, M.; Poelman, M.; Deflorian, F.; Rossi, S. Compatibility between cataphoretic electro-coating and silane surface layer for the corrosion protection of galvanized steel. *Prog. Org. Coat.* **2010**, *69*, 118–125. [CrossRef]
20. Romano, A.P.; Fedel, M.; Deflorian, F.; Olivier, M.G. Silane sol-gel film as pretreatment for improvement of barrier properties and filiform corrosion resistance of 6016 aluminium alloy covered by cataphoretic coating. *Prog. Org. Coat.* **2011**, *72*, 695–702. [CrossRef]
21. Manas, D.; Manas, M.; Gajzlerova, L.; Ovsik, M.; Kratky, P.; Senkerik, V.; Skrobak, A.; Danek, M.; Manas, M. Effect of low doses beta irradiation on micromechanical properties of surface layer of injection molded polypropylene composite. *Radiat. Phys. Chem.* **2015**, *114*, 25–30. [CrossRef]
22. Mishra, R.; Behera, B.K.; Muller, M.; Petru, M. Finite element modeling based thermodynamic simulation of aerogel embedded nonwoven thermal insulation material. *Int. J. Therm. Sci.* **2021**, *164*, 106898. [CrossRef]
23. Botko, F.; Hatala, M.; Beraxa, P.; Duplak, J.; Zajac, J. Determination of CVD Coating Thickness for Shaped Surface Tool. *TEM J. Technol. Educ. Manag. Inform.* **2018**, *7*, 428–432. [CrossRef]
24. Jurko, J.; Panda, A.; Gajdos, M.; Zaborowski, T. Verification of Cutting Zone Machinability during the Turning of a New Austenitic Stainless Steel. *Adv. Comput. Sci. Educ. Appl.* **2011**, *202*, 338.
25. Behalek, L.; Novak, J.; Brdlik, P.; Boruvka, M.; Habr, J.; Lenfeld, P. Physical Properties and Non-Isothermal Crystallisation Kinetics of Primary Mechanically Recycled Poly(l-lactic acid) and Poly(3-hydroxybutyrate-co-3-hydroxyvalerate). *Polymers* **2021**, *13*, 3396. [CrossRef] [PubMed]
26. Svetlik, J.; Malega, P.; Rudy, V.; Rusnak, J.; Kovac, J. Application of Innovative Methods of Predictive Control in Projects Involving Intelligent Steel Processing Production Systems. *Materials* **2021**, *14*, 1641. [CrossRef] [PubMed]
27. Milosevic, M.; Cep, R.; Cepova, L.; Lukic, D.; Antic, A.; Djurdjev, M. A Hybrid Grey Wolf Optimizer for Process Planning Optimization with Precedence Constraints. *Materials* **2021**, *14*, 7360. [CrossRef] [PubMed]
28. Nova, I.; Frana, K.; Solfronk, P.; Sobotka, J.; Korecek, D.; Svec, M. Characteristics of Porous Aluminium Materials Produced by Pressing Sodium Chloride into Their Melts. *Materials* **2021**, *14*, 4809. [CrossRef]
29. Baron, P.; Kocisko, M.; Blasko, L.; Szentivanyi, P. Verification of the operating condition of stationary industrial gearbox through analysis of dynamic signal, measured on the pinion bearing housing. *Measurement* **2017**, *96*, 24–33. [CrossRef]
30. Panda, A.; Duplak, J.; Jurko, J.; Behun, M. New experimental expression of durability dependence for ceramic cutting tool. *Appl. Mech. Mater.* **2013**, *275–277*, 2230–2236. [CrossRef]
31. Cuha, D.; Hatala, M. Effect of a modified impact angle of an ultrasonically generated pulsating water jet on aluminum alloy erosion using upward and downward stair trajectory. *Wear* **2022**, *500*, 204369. [CrossRef]
32. Bukovska, S.; Moravec, J.; Solfronk, P.; Pekarek, M. Assessment of the Effect of Residual Stresses Arising in the HAZ of Welds on the Fatigue Life of S700MC Steel. *Metals* **2022**, *12*, 1890. [CrossRef]
33. Kamble, Z.; Mishra, R.K.; Behera, B.K.; Tichy, M.; Kolar, V.; Muller, M. Design, Development, and Characterization of Advanced Textile Structural Hollow Composites. *Polymers* **2021**, *13*, 3535. [CrossRef]

Disclaimer/Publisher's Note: The statements, opinions and data contained in all publications are solely those of the individual author(s) and contributor(s) and not of MDPI and/or the editor(s). MDPI and/or the editor(s) disclaim responsibility for any injury to people or property resulting from any ideas, methods, instructions or products referred to in the content.

Article

Investigation of Oxide Thickness on Technical Aluminium Alloys—A Comparison of Characterization Methods

Ralph Gruber [1], Tanja Denise Singewald [1], Thomas Maximilian Bruckner [1], Laura Hader-Kregl [1], Martina Hafner [2], Heiko Groiss [3], Jiri Duchoslav [4] and David Stifter [4,*]

[1] CEST—Centre for Electrochemistry and Surface Technology, Viktor-Kaplan Str. 2, 2700 Wiener Neustadt, Austria; ralph.gruber@cest.at (R.G.); tanja.singewald@cest.at (T.D.S.); thomas.bruckner@cest.at (T.M.B.); laura.hader-kregl@cest.at (L.H.-K.)
[2] AMAG Rolling GmbH, Lamprechtshausener Str. 61, 5282 Ranshofen, Austria; martina.hafner@amag.at
[3] Christian Doppler Laboratory for Nanoscale Phase Transformations, Center for Surface and Nanoanalytics, Johannes Kepler University Linz, Altenberger Str. 69, 4040 Linz, Austria; heiko.groiss@jku.at
[4] ZONA—Center for Surface and Nanoanalytics, Johannes Kepler University Linz, Altenberger Str. 69, 4040 Linz, Austria; jiri.duchoslav@jku.at
* Correspondence: david.stifter@jku.at

Citation: Gruber, R.; Singewald, T.D.; Bruckner, T.M.; Hader-Kregl, L.; Hafner, M.; Groiss, H.; Duchoslav, J.; Stifter, D. Investigation of Oxide Thickness on Technical Aluminium Alloys—A Comparison of Characterization Methods. *Metals* 2023, *13*, 1322. https://doi.org/10.3390/met13071322

Academic Editor: Wangping Wu

Received: 3 July 2023
Revised: 20 July 2023
Accepted: 21 July 2023
Published: 24 July 2023

Copyright: © 2023 by the authors. Licensee MDPI, Basel, Switzerland. This article is an open access article distributed under the terms and conditions of the Creative Commons Attribution (CC BY) license (https:// creativecommons.org/licenses/by/ 4.0/).

Abstract: In this study the oxide layer of technical 6xxx aluminium surfaces, pickled as well as passivated, were comparatively investigated by means of transmission electron microscopy (TEM), Auger electron and X-ray photoelectron spectroscopy (AES, XPS), the latter in two different operating modes, standard and angle resolved mode. In addition, confocal microscopy and focused ion beam cutting were used for structural studies of the surfaces and for specimen preparation. The results illustrate in detail the strengths and weaknesses of each measurement technique. TEM offers a direct way to reliably quantify the thickness of the oxide layer, which is in the range of 5 nm, however, on a laterally restricted area of the surface. In comparison, for AES, the destructiveness of the electron beam did not allow to achieve comparable results for the thickness determination. XPS was proven to be the most reliable method to reproducibly quantify the average oxide thickness. By evaluating the angle resolved XPS data, additional information on the average depth distribution of the individual elements on the surface could be obtained. The findings obtained in this study were then successfully used for the investigation of the increase in the aluminium oxide thickness on technical samples during an aging test of 12 months under standard storage conditions.

Keywords: technical aluminium alloy; aluminium oxide thickness; oxide thickness characterization; method comparison; X-ray photoelectron spectroscopy; Auger electron spectroscopy; transmission electron microscopy; oxide growth; aging test

1. Introduction

Due to global CO_2 emission regulations and the general phase-out of fossil fuels, car manufacturers tend to use more and more lightweight materials to fulfil the required environmental objectives. The change to lighter materials helps to reduce fuel consumption and gives new opportunities in technical applications and design concepts [1–3]. Over the last decades aluminium has been one of the most promising materials for the automotive industry. With a material density about 65% lower than steel, the main advantage is to be found in weight reduction [4]. However, aluminium—especially high-strength alloys—also satisfies the torsion and stiffness requirements for automotive applications and provides outstanding shock absorbing properties which helps to increase passenger safety [2].

Based on the above introduced reasons, structural adhesive bonding for joining car body parts has been established by the automotive industry [5]. This bonding technique, also in combination with other joining methodologies, shows significant advantages in crash performance, cost effectiveness and in the ability of multi-material approaches in lightweight constructions of automobiles [6]. The most widely used types of structural adhesives for automotive applications are epoxies. These cross-linked polymers are bonding to the aluminium surface and build strong adhesive joints. The long-term durability of those is highly dependent on the surface chemistry of the aluminium substrate prior to bonding—hence the characterization of the top surface layers is of considerable technological importance [4]. Especially, the thin amorphous γ-Al_2O_3 phase on the surface of aluminium and its alloys has a significant effect on the corrosion resistance, wettability and adhesion performance and therefore has often been studied for a better understanding of the relationship between composition and thickness of the aluminium oxide layer and the resulting bonding performance [7–11].

Routinely, oxide thicknesses can be determined using closely industry-related measurement techniques, in particular glow discharge optical emission spectroscopy (GDOES) or infrared reflection absorption spectroscopy (IRRAS). Both methods are established analysis methods for oxide layer thicknesses and are mainly used where corresponding layer thicknesses over 100 nm are required [12]. However, a comparison carried out by us in advance of the current study showed that both techniques, despite being well established methodologies in their field of application, are unsuitable for a reliable determination of aluminium oxide thicknesses below 10 nm.

A technique to characterize aluminium oxide layers in the range of only a few nanometres in thickness is X-ray photoelectron spectroscopy (XPS). The calculation of the oxide thickness can be performed by considering an equation proposed by Strohmeier, which uses the ratio of the measured Al 2p oxide to metal peak intensities [13]. Providing a similar information depth as XPS (i.e., approximately 10 nm), Auger electron spectroscopy (AES) can also be applied for the oxide thickness determination using the evaluation according to Strohmeier due to a comparable energy range of the electrons. In contrast, the quantification of aluminium oxide thicknesses by high resolution images using transmission electron microscopy (TEM) is often performed on a highly localized scale to provide a validation of the values obtained with other characterization techniques [14–16].

Considering the significant effect of an aluminium oxide surface layer on different application properties and the importance of the exact compositional characterization, the question arises, which of the above-mentioned techniques is a valid method for oxide thickness determination on industrial 6xxx aluminium alloys. For this purpose, a comparison was carried out by means of XPS, AES and TEM. Moreover, XPS was taken to quantify the surface chemistry of the aluminium substrates in detail, while TEM was used to describe the top surface layer structure. The resulting oxide thicknesses show a dependency on the chosen method. Consequently, it is necessary to weigh up the advantages and disadvantages of each method for a meaningful characterization of the surface oxide layer.

The findings were then applied to reliably determine the thickness increase in oxide layers on technical aluminium surfaces in an aging study by storing the samples over a period of 12 months under ambient conditions.

2. Materials and Methods
2.1. Materials

As representatives for technical aluminium surfaces used by the automotive industry, an AA 6016 AlMgSi alloy with two different commercial surface finishes was provided by AMAG Austria Metall AG (Ranshofen, Austria). The primary surface treatment was based on automotive standards. In detail, after the finishing annealing, the substrate surface was cleaned by means of acid pickling and a subsequent washing process. After this, the surface was textured using electrical discharge texturing (EDT). A second type of sample experienced a further passivation step with a commercial conversion coating after the cleaning and before surface texturing. Table 1 gives an overview of the sample types used (A1, A2) with the differently applied surface treatments.

Table 1. Overview of investigated sample types and their different surface treatments.

Label	Alloy	Composition	Surface Treatment Procedure	Surface Texture
A1	AA6016	Al98Mg0.6Si1.4	pickling and cleaning	EDT
A2	AA6016	Al98Mg0.6Si1.4	pickling, cleaning and passivation	EDT

2.2. Sample Preparation

Using a metal plate shear the aluminium samples were cut down into pieces with dimensions of 6×6 mm^2, 12×12 mm^2 or 20×20 mm^2, depending on the used characterization method. Prior to analysis, the surfaces of the specimens were cleaned by a three-step ultrasonic degreasing procedure, using different organic solvents, namely aceton ($\geq$99.0% Ph. Eur., VWR Chemicals, Radnor, IN, USA), tetrahydrofurane (THF, $\geq$99.0% ACS reagent grade, contains 250 ppm BHT as inhibitor, Sigma-Aldrich, St. Louis, MO, USA) and isopropanol ($\geq$99.5% ACS reagent grade, Sigma-Aldrich, St. Louis, MO, USA). The samples were put into a 100 mL beaker, filled with approximately 40 mL of one of the three different solvents and put into an ultrasonic bath for 30 min each.

2.3. Instrumentation

As the analysis method with highest lateral resolution in this comparative study, TEM provides the means to directly measure the oxide thickness in a quantitative way, however, confined to a rather small local area. Therefore, the TEM results were used as reference for the thickness evaluation by choosing representative regions on the sample surfaces. For that purpose, focused ion beam (FIB) cutting in a 1540XB Crossbeam system (Zeiss, Oberkochen, Germany) was used as a preparation method to obtain appropriate specimens for TEM analysis. With gallium ions at 30 keV, lamellae were cut out of the sample surfaces and afterwards thinned to electron transparency. Prior to FIB preparation the surface of the samples were protected by an electron stimulated platinum deposition to avoid damaging the oxide layer during the subsequent preparation steps. The positions from where the lamellae were cut out were chosen after an investigation of the surface structure, roughness and texture using a MarSurf CM mobile confocal microscope (Mahr GmbH, Göttingen, Germany). To measure the oxide layer thickness, TEM measurements were performed in a JEM-2200FS system (JEOL, Akishima, Japan), using an accelerating voltage of 200 kV. The high-resolution TEM images were recorded by means of zero-loss filtering, using the in-column Ω-filter. The TEM system is additionally equipped with an energy-dispersive X-ray detector (EDX) from Oxford Instruments (Abingdon, UK) for elemental analysis using the microscope in the scanning mode (STEM).

Providing different measurement modes, XPS was chosen to be a much faster and more versatile method which provides complementary data. Next to an overview of the elemental and chemical composition of the surface layers, XPS was mainly used to obtain the oxide thickness. All measurements were performed using a Thetaprobe XPS system (Thermo Scientific, Waltham, MA, USA). The device features a monochromatic Al K$_\alpha$ X-ray source ($h\nu$ = 1486.6 eV) and is equipped with a dual flood gun for surface charge neutralization. The spot size of the X-ray beam was chosen to be 400 µm in diameter in order to obtain an average oxide layer thickness. To exploit the whole data potential of this analysis technique the XPS measurements were performed in standard as well as in angle resolved mode. For each mode, survey spectra were obtained using 200 eV as pass energy with an energy step width of 1 eV. More detailed high resolution (HR) spectra were recorded with a pass energy of 50 eV and 0.05 eV step width. The evaluation of the XPS data was performed using the Avantage software package from the system supplier.

As AES is also sensitive to the same surface near region as XPS, a scanning AES microscope (JEOL JAMP 9500F, Akishima, Japan) was additionally used to determine the oxide layer thickness. For spectroscopy of Auger electrons, the device is equipped with a hemispherical electron energy analyser and a channeltron detector, positioned at an angle of 60° with respect to the beam direction of the electron gun. As the AES microscope can record images using secondary and backscatter electron detectors the system was also applied to determine the microstructure of the textured aluminium surface. The investigations were carried out with the sample surface perpendicular to the electron beam using 10 kV accelerating voltage and a beam current of 10 nA.

3. Results

For the comparison of the different measurement techniques with respect to their performance in obtaining the oxide thickness, it was necessary to define a reference value for each sample. The commercial aluminium alloy with the two different types of surface treatment was investigated, at first, by means of TEM. Due to the complex EDT pattern on the surfaces the positioning for the FIB preparation was crucial to quantify the oxide thickness in a representative way for the (majority of) the surface. Figure 1 shows the EDT structured surface of the pickled aluminium sheet of type A1 observed by confocal microscopy.

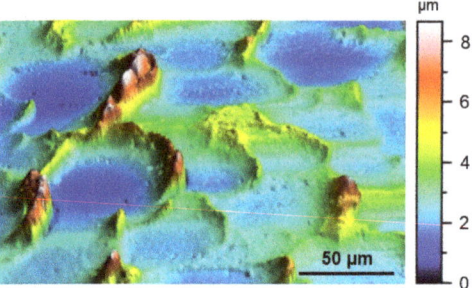

Figure 1. EDT surface structure of a pickled specimen of type A1 investigated by confocal microscopy.

The imprints result from the rolling process and lead to an average surface roughness of $R_a = 0.65$ µm. Due to high shear forces during the rolling process the edges of the EDT craters often show distinctly different microstructure compared to the flatter regions, which account for the largest portion of the surface area. In addition, rolling and texturing affects the local surface properties due to the mechanical displacement during the production procedure and is often described as a surface near deformed layer (NSDL) [17–19]. As shown in Figure 2 the positions for the FIB preparation of the lamellae were chosen to be in a flatter and homogeneous area of the sample. This guarantees a rather uniform oxide layer without influences of the extreme surface texture or the NSDL on the ridges caused by the production process.

Figure 2. Position and preparation steps of a TEM lamella using FIB cutting and thinning.

3.1. Reference Value from Transmission Electron Microscopy

To obtain representative values for a comparison of the methods the oxide layer thickness was measured from TEM images similar to the one presented in Figure 3. A section of a lamella of sample type A1 with a clearly visible uniform oxide layer on top of the aluminium bulk is shown. The darker area in the image represents the sputter-coated platinum layer which avoided damage to the oxide layer during FIB preparation. It can be clearly differentiated from the brighter and grey coloured aluminium oxide layer. For better statistical and representative oxide thickness values, the measurement was carried out at least five times at different sections on a lamella for each sample type.

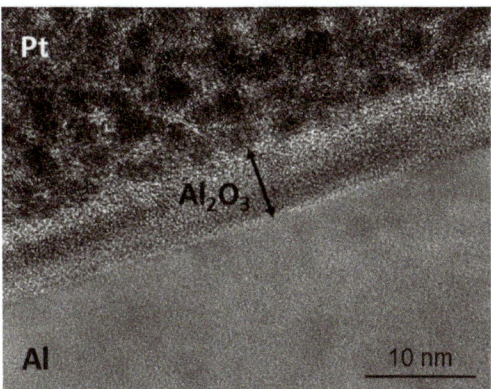

Figure 3. Presence of a uniform oxide layer on a subsection of a lamella of type A1.

Since the production process of the two sample types differs only by one additional passivation step without any further rise in temperature, the expected oxide layer thicknesses for both types should be in the same nanometre range. The obtained average oxide layer thicknesses for both types, given in Table 2, confirms this assumption.

Table 2. TEM results of the oxide layer thicknesses.

Sample	Av. Oxide Layer Thickness [nm]
A1	5.8 ± 0.6
A2	5.6 ± 0.3

EDX-mappings were additionally performed on sections of the lamellae, exhibiting a difference in the layer structure between the pickled and the additional passivated sample types. As shown in Figure 4, sample type A1 differs from A2 due to the presence of a non-uniform conversion layer. The aluminium bulk (light blue) is fully covered by an oxide layer (green). For A2 titanium (red), which is part of the conversion coating, is visibly deposited on top of the oxide in a non-uniform way. The peninsular-like growth of the zirconium and titanium-based conversion layer is electrochemically driven. In detail, during the initial stages of the coating process the oxide layer experiences a dissolution due to the fluoride containing coating solution. The surface near intermetallic phases becomes more cathodic, which in turn favours the formation of a conversion layer in the region surrounding those phases [20].

3.2. XPS

After the fundamental characterization by TEM, all sample types were investigated using XPS in two different operation modes, standard as well as angle resolved, for oxide thickness determination. Compared to the spatially restricted local analysis provided by TEM, XPS offers the advantage of being a much faster and laterally averaging method which is more suitable for the characterization of technical sample surfaces. Consequently, the X-ray beam size for both analysis modes was chosen to be 400 µm in diameter. In addition, XPS does not require tedious sample preparation steps prior analysis and is nearly non-destructive, in contrast to AES, as presented later in this study [21].

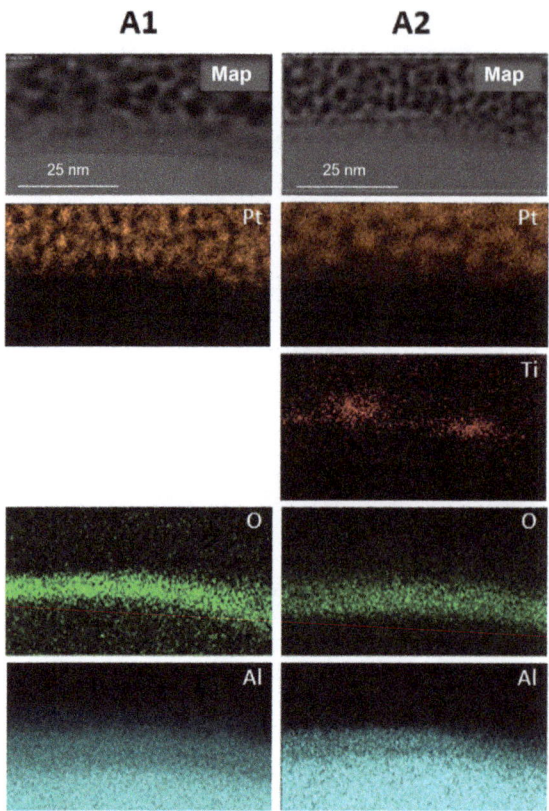

Figure 4. Layer structure comparison of both sample types by EDX elemental mapping.

At first, standard mode XPS was used to obtain an overview of the elemental composition of the topmost surface layer of all surface treated aluminium samples. The comparison of the survey spectra shown in Figure 5 reveals the difference of the surface chemistry due to the different pre-treatment procedures. Despite the three-step cleaning procedure with organic solvents, all samples revealed carbon contamination with a concentration around 30 at.% on the surfaces. In addition, also other contamination elements, e.g., Na, S and Ca could be detected on the analysed surfaces in negligible traces under 1 at.%. By omitting these contaminations, a general elemental surface composition could be evaluated as summarized in Table 3. Comparing the major constituents of the samples, it was found that Al, O and C are the dominating elements present on the surfaces, closely followed by F with varying concentrations, depending on the pre-treatment process. The high carbon contamination content at the samples surface, partly bound to oxygen, could be a reason for the difference in the determined stoichiometry of the Al_2O_3 oxide, as shown in Table 3. The fluorine concentration of the non-passivated aluminium alloy of type A1 can be attributed to the fluorine-containing ingredients of the pickling bath. Most of the detected fluorine as well as all of titanium and zirconium on the passivated sample type A2 originate from the conversion coating.

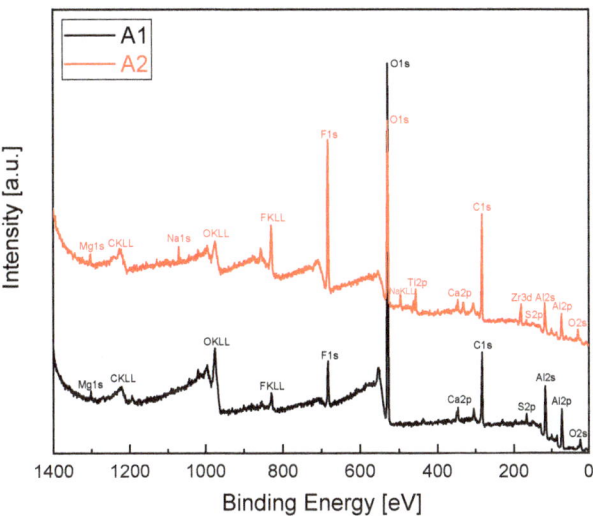

Figure 5. Comparison of XPS survey spectra of both sample types.

Table 3. Elemental composition of the surface layer on all analysed samples obtained by XPS.

Sample	Elemental Composition [at.%]						
	Al	O	C	F	Mg	Ti	Zr
A1	19	38	31	5	0.4		
A2	16	29	32	18	0.6	1.4	1.1

To better identify the chemical components in the topmost surface layers, HR spectra of the major elemental constituents, aluminium, oxygen, fluorine and carbon for sample type A1, and additional titanium and zirconium for sample type A2, were additionally recorded. From the results of the HR data, it was derived that the sample surfaces are homogeneous in terms of chemical composition and only differ due to their very own pre-treatment history. The bulk aluminium alloy, which predominantly consists of aluminium without any significant alloying element contents in the topmost surface areas, is covered by an Al_2O_3 layer on all samples. Pickled specimens show small amounts of fluorine, which originates from components of the pickling bath solution, embedded into the oxide layer. On the passivated samples fluorine could equally be detected within the oxide layer, although, in much higher concentrations. In combination with titanium and zirconium on top of the oxide this result, as mentioned above, is attributed to the conversion coating process during production. As already seen from the survey spectra, carbon was detectable on both surfaces despite an intense cleaning procedure. By fitting the high-resolution spectrum of the C1s core level three different carbon peaks could be identified (C-C/H, C-O, C=O), which are mostly attributed to different components of the lubricants which are used during the rolling process of the aluminium sheets for automotive applications.

The above findings are based on HR spectra, which were taken in the angle resolved mode of the XPS. This AR data was used, in addition to the identification of chemical composition in the topmost surface layers, to also qualitatively estimate the depth distribution of the major elemental constituents. By taking the negative logarithm of the peak intensity ratio collected at bulk- and surface-sensitive angles, as shown in Figure 6, an average relative depth position for each element can be obtained. These relative depth plots do not have a quantitative depth-axis since the absolute depth position is not derived by this method. However, these plots qualitatively indicate the order of the different elemental species with respect to each other [22,23].

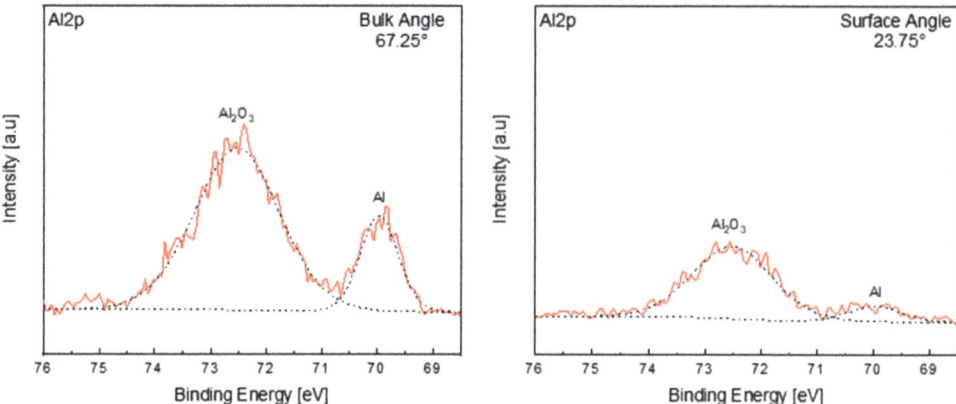

Figure 6. ARXPS surface and bulk angle HR spectra of Al2p.

In Figure 7a the relative depth position of each surface related element of sample type A2 is shown. The carbon contamination originates from the topmost surface region, whereas the signals of the major elemental constituents of the conversion coating, i.e., Zr, Ti, F and O, come from the intermediate information depth range of this method. The aluminium oxide layer is also found in the middle depth region directly beneath the conversion coating. The signals of alloying elements such as aluminium, magnesium and silicon originate from larger depths within the total information depth range of ARXPS and are allocated towards the bulk section of the sample. In addition to the determination of relative depth positions of different elements, ARXPS provides the means to obtain even full in-depth concentration profiles in a non-destructive way. In Figure 7b the corresponding virtual depth profile from sample type A2 is depicted. The findings reflect the results of the relative depth plot and confirm the layered structure as observed in the TEM analysis. The elements originating from the passivation step are located on top of the aluminium oxide layer and are covered by carbon contaminations. Alloying elements such as magnesium and silicon can be found again towards the aluminium bulk section.

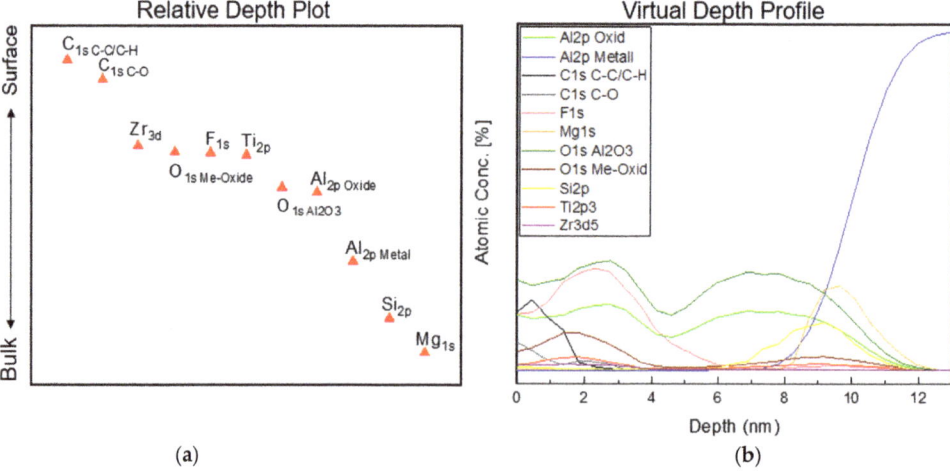

Figure 7. (**a**) Relative depth plot (**b**) Corresponding virtual depth profile. Both obtained by ARXPS.

Next to an overview of the elemental and chemical composition of the surface layers, XPS was mainly used to obtain the oxide thickness. The calculation method is based on the

Beer-Lambert law and has been introduced by Strohmeier [13]. The equation, shown in Equation (1), covers the ratio of the measured Al2p oxide to metal peak intensities with respect to the elemental inelastic mean free paths (IMFP) λ_m and λ_o, the atomic volume densities N_m and N_o as well as the average take off angle θ and can be applied for true oxide layer thicknesses < 10 nm.

$$d_{oxide}(\text{Å}) = \lambda_o \sin\theta \ln\left(\frac{N_m \lambda_m I_o}{N_o \lambda_o I_m} + 1\right) \quad (1)$$

The volume density ratio of aluminium atoms in metal to oxide used in this study was $N_m/N_o = 1.5$. Furthermore, the average take off angle was $\theta = 60°$ and for the IMFP values in metallic aluminium $\lambda_m = 26$ Å and in Al_2O_3 $\lambda_o = 28$ Å was used [13,24]. In consideration of the given parameters the aluminium oxide thickness was calculated using the Al2p HR spectra from both, standard as well as angle resolved mode. The corresponding intensities I_m and I_o were obtained by fitting the peak with a higher binding energy as Al_2O_3 (Al2p oxide) and the other at lower binding energy as metallic aluminium (Al2p metal). Compared to the values obtained by TEM, the results of the standard XPS measurement in Table 4 are close, but show slightly lower oxide thicknesses. The underestimation of the thickness is, besides in the limitation of the Strohmeier model, due to the fact that the chosen spot size of the X-ray beam is averaging the results in the observation area. Aside from that of the XPS technique itself, in angle resolved or standard mode, it has little to no effect on the results when performing the thickness calculation based on the Strohmeier equation. This can be explained by the evaluation procedure. In order to obtain the corresponding Al2p spectra out of the AR data, one has to collapse every spectrum from each measured angle to obtain one single main spectrum. This main Al2p spectrum differs hardly from the spectrum as obtained by standard mode XPS measurements.

Table 4. Results of oxide thickness evaluation by different XPS-based methods.

Sample	Oxide Thickness Standard Mode [nm]	Oxide Thickness AR-Overlayer [nm]	Oxide Thickness Virtual Depth Profile [nm]
A1	4.7 ± 0.1	4.1 ± 0.1	6.2 ± 0.6
A2	4.9 ± 0.2	4.3 ± 0.2	6.5 ± 0.8

Aside from the oxide thickness calculation based on the procedure proposed by Strohmeier, ARXPS provides an overlayer thickness calculation procedure implemented in the Avantage software package of the system manufacturer. The calculation is based on all of the generated AR data (and not only on a collapsed single spectrum) and requires additional information about material properties, i.e., chemical formula, density and bandgap in order to obtain the corresponding IMFPs for the thickness calculation. The results of this evaluation, taking the single HR AR spectra into account, are given in Table 4 (named AR-Overlayer).

A further possibility for the determination of the oxide thickness is based on virtual depth profiles from ARXPS data as shown in Figure 7 for sample type A2. The oxide thickness can be derived from the Al2p oxide signal with respect to the reconstructed depth-axis. The results of this evaluation are also shown in Table 4. The obtained values indicate that the depth axes in virtual depth profiles from ARXPS has to be taken with care and that an analysis of the oxide thickness in a quantitative way is not straight forward, especially on technical aluminium alloys. However, such an AR-based evaluation complements the elemental and chemical surface characterization by standard XPS analysis and is a valuable technique to reveal the existing layer structure on industrial aluminium surfaces.

3.3. AES

Providing a similar information depth as XPS from the same surface near region AES was chosen to be the third method for this comparative study. Due to the same range of electron energies, identical IMFP values for aluminium and its oxide as for XPS can be used and the determination of the oxide thickness using the equation after Strohmeier (Equation (1)) is applicable. Consequently, all samples were evaluated using the first derivative of the fitted high resolution aluminium spectra, such as the one shown in Figure 8. The main advantage of using the first derivative of measured AES spectra is the background suppression. The ratio of the measured aluminium oxide to metal peak intensities, which are necessary for the application of Equation (1), were calculated using the difference of the minimum and maximum points from each compound peak structure. All spectra were fitted equally by means of a linear combination fit. For that purpose, acquired spectra of pure reference materials, metallic aluminium and Al_2O_3, were used in combination to fit the spectra of the samples. This procedure generates a reproduceable way to qualitatively evaluate high resolution AES measurements.

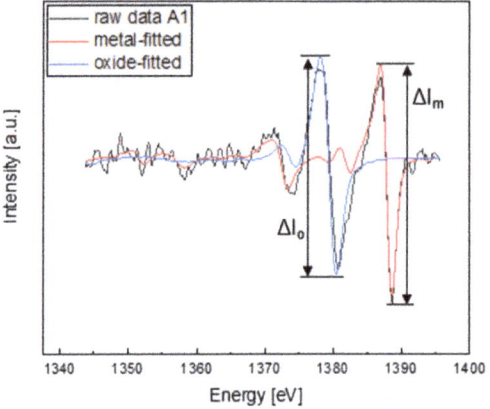

Figure 8. AES high resolution Al spectrum with fitted components.

As mentioned above, the usage of focused electron beams can lead to a degradation of metallic oxides [21]. This observation was also made in the current study when performing point measurements on all samples. As shown in Figure 9, HR spectra were additionally recorded using area measurements instead of point measurements in order to decrease degradation effects. By stepwise doubling the field of investigation the influence of different measurement areas on the oxide thickness values was obtainable. The areas were positioned in the flatter zones of the EDT surface to reliably quantify the oxide thickness. The findings given in Table 5 clearly demonstrate the degradation effect of the electron beam depending upon the resulting beam exposure. The average oxide thickness of sample type A1 obtained using point measurements is more than three times less than the average acquired by area investigations. The increase in the measured area also involves an increase in the thickness of the investigated oxide layer. Comparing a measurement area four times the size as the initial one, the increase in oxide thickness is approximately 46%.

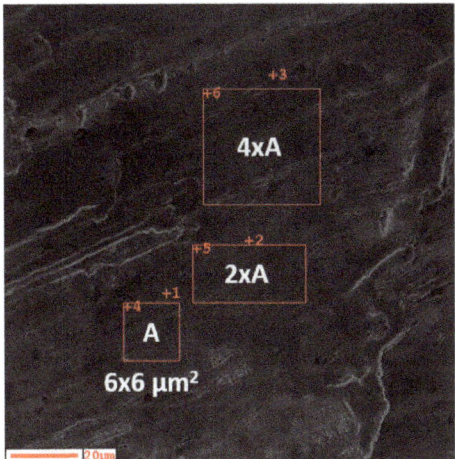

Figure 9. Observation area during AES measurements.

Table 5. Comparison of obtained thickness values for point and area AES measurements.

Sample	Measurement Point/Area	Oxide Thickness [nm]	Mode
A1	1–3	0.69	Point
	4	2.11	Area A
	5	2.60	Area 2A
	6	3.10	Area 4A

4. Discussion

To further elucidate which method is the most practical one for a reliable determination of the oxide thickness on technical aluminium alloys, the overall results are compared in Figure 10. TEM provides the highest lateral resolution. Although this method offers the best way to reliably quantify the oxide thickness, it should be considered that the results are referred only to a very confined region of the real samples surface. To obtain statistical validity of the measurement data, more different measurement positions evenly distributed on the surface are recommended to draw reliable conclusions about the average oxide thickness of the investigated samples. TEM is also not practicable for quick analysis due to laborious sample preparation procedures. Nevertheless, the results of the TEM analysis in this study show that all investigated sample types have a similar oxide thickness despite different treatment processes, namely the additional passivation step of sample type A2 has no significant effect on the thickness of the initial oxide layer.

Comparing the results of the two different modes for XPS measurements, oxide thickness values from the standard mode differ only slightly from the values obtained by AR mode. However, the thickness determination from virtual depth profiles seems to show significantly higher values as compared to the so-called overlayer model and the calculation according to Strohmeier. This discrepancy can be explained by inaccuracies of input parameters needed for generating such virtual depth profiles, as well as by limitations of the underlying model for reconstructing the profile out of the ARXPS data. There are several factors, e.g., the indication of the correct chemical composition of the different surface compounds, which have an influence on the results. Nevertheless, by providing the opportunity to obtain information about elemental and chemical surface composition and additionally about the average depth distribution of different elemental species and the layer structure (e.g., in relative depth plots), the angle resolved mode is generally preferred. Comparing the XPS results to the TEM ones, a small discrepancy in the range of less than a nanometre occurs. This difference is attributed to a combination of several factors. Firstly, standard curve fitting of the metallic Al2p main peak was proven to have an error in the percentage range. This discrepancy originates from an overestimation of contributed substrate photoelectrons to the inelastic background [24]. Furthermore, XPS is a much more averaging method as compared to the highly localized investigation by TEM. Since the investigated samples originate from industrial production and exhibit a surface texture with a certain roughness, local oxide thickness deviations may exist. In addition, the TEM lamellae, in the size of approximately 20 µm, were prepared from the flatter areas of the EDT imprints, giving values without influences from the surface texture, especially from the ridges and the deformed microstructure at the edges of the EDT imprints [17–19,24]. With an average spot size of 400 µm in diameter XPS is sensitive to every part of the textured aluminium surface. Hence, the values of the oxide thickness obtained by XPS may give the most relevant overall picture for technical aluminium surfaces.

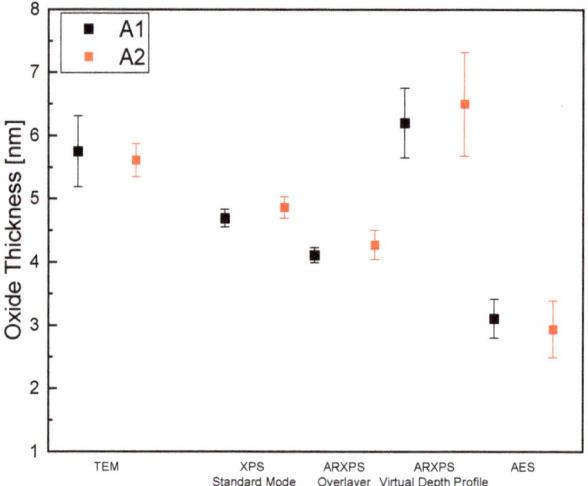

Figure 10. Comparison of obtained Al_2O_3 thickness values by different measurement methods.

Finally, the destructiveness of the focused electron beam during AES measurement becomes apparent in the results. Especially point measurements show a significant degradation of the oxide layer. Thus, area measurements were chosen to suppress and overcome degradation to a certain degree. The smallest possible area for the given magnification was 6×6 µm^2, which led to a less destructive interaction with the oxidic surface. Hence, the measured oxide thickness increased by a factor of three with respect to point measurements. Nevertheless, the resulting oxide layer still deviates significantly from the TEM and XPS results. Therefore, the measurement area was increased stepwise up to 12×12 µm^2, resulting in an improvement, but being still far below the values obtained by the other methods. Nevertheless, AES can act as a link between the nanoscopic view of TEM and the laterally averaging one of XPS, when taking degradation into account.

Based on our findings, the increase in the aluminium oxide thickness over time was observed during aging for one year at standard storage conditions. The results obtained by XPS were compared again with those by TEM, since these methods are the most accurate ones, and are shown in Figure 11. As known from the literature, the growth rate of aluminium oxide at ambient pressures and temperatures is initially extremely rapid and decreases with increasing thickness of the oxide. In general, the growth rate is positively influenced by increased temperatures to a certain point and elevated oxygen concentrations [25]. However, as soon as a so-called limiting thickness is reached, the oxide growth stops and does not change with time [26]. The aluminium oxide of the investigated 6xxx alloy also follows this trend. It seems that the limiting thickness is reached already before half a year of aging for both sample types. It shall be noted that the initial difference in the obtained oxide thickness between the two analysis methods, XPS and TEM, could also be detected again after one year of aging for both samples. Of especial importance is the gained knowledge that XPS underestimates TEM data by approximately one nanometre. Consequently, aluminium oxide layer determination by means of XPS is a reproducible and representative, as well as a relatively easily to be performed method for the characterization of technical aluminium surfaces.

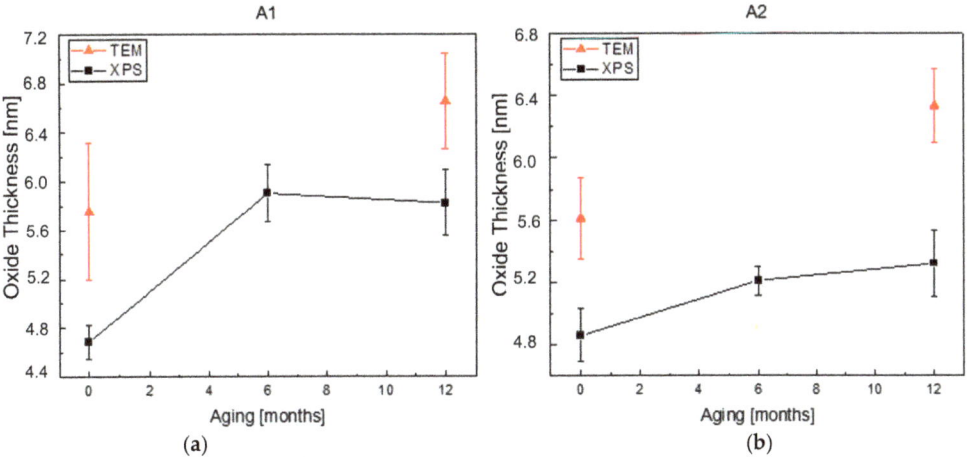

Figure 11. Increase in the aluminium oxide thickness over time during aging at standard storage conditions for one year observed by means of XPS and TEM. (**a**) shows results of sample type A1, (**b**) refers to sample A2.

5. Conclusions

In this comparative study, a valid and reliable method for oxide thickness determination on an industrial 6xxx aluminium alloy was searched for. The oxide thicknesses were determined by means of TEM, XPS—in two different operating modes, standard and angle resolved mode—as well as by AES.

(1) The results show that XPS gives reproducibly accurate averaged values of the thickness and provides even the possibility to generate information about elemental and chemical surface composition and additionally provides information about the average depth distribution of different elemental species and the resulting layer structure.

(2) Although TEM offers the highest lateral resolution and therefore the most direct and accurate way for oxide thickness quantification, it should be considered that the results are referred only to a rather confined region of the real samples surface. Reliable conclusions based on TEM measurements require statistical validity by accessing multiple measurement positions evenly distributed on the samples surface. Furthermore, sample preparation for TEM investigations requires significantly higher experimental effort as compared, e.g., for XPS.

(3) In this study, AES acted as a link between the nanoscopic view of TEM and the laterally averaging characteristics of XPS. However, the destructiveness of the focused electron beam during AES measurements becomes apparent in the results, where a significant degradation of the oxide layer was detected.

(4) Finally, an aging test at standard storage conditions proved that the aluminium oxide layer determination by means of XPS is reproducible and in combination with a short measurement time and minimal sample preparation requirements, a highly suitable method for the characterization of technical aluminium surfaces. This work further clearly illustrates the importance of knowing the individual characteristics

and strengths of each measurement technique, especially when performing industry related applied research.

Author Contributions: Conceptualization, R.G. and D.S.; methodology, D.S. and H.G.; formal analysis, J.D., T.D.S. and T.M.B.; investigation, R.G.; resources, M.H. and D.S.; data curation, R.G.; writing—original draft preparation, R.G.; writing—review and editing, D.S.; supervision, L.H.-K., M.H. and D.S.; project administration, L.H.-K. and M.H.; All authors have read and agreed to the published version of the manuscript.

Funding: The Comet Centre CEST is funded within the framework of COMET—Competence Centers for Excellent Technologies by BMVIT, BMDW as well as the Province of Lower Austria and Upper Austria. The COMET programme is run by FFG. This work originates from research in the Durabond and GreenMetalCoat (FFG 865864, Comet-Center (K1), 2019–2022; FFG 899594, Comet-Center (K1), 2023) projects.

Data Availability Statement: Not applicable.

Acknowledgments: J.D. and D.S. acknowledge the government of Upper Austria for financial support (project ASAES). The financial support by the Austrian Federal Ministy of Labour and Economy, the National Foundation of Research, Technology and Development and the Christian Doppler Research Association is gratefully acknowledged.

Conflicts of Interest: The authors declare no conflict of interest.

References

1. Njuguna, J. *Lightweight Composite Structures in Transport: Design, Manufacturing, Analysis and Performance*; Elsevier Science: Amsterdam, The Netherlands, 2016.
2. Hirsch, J. Recent development in aluminium for automotive applications. *Trans. Nonferrous Met. Soc. China* **2014**, *24*, 1995–2002. [CrossRef]
3. Crolla, D.; Ribbens, W.; Heisler, H.; Blundell, M.; Harty, D.; Brown, J.; Serpento, S.; Robertson, A.; Garrett, T.; Fenton, J.; et al. *Automotive Engineering e-Mega Reference*; Elsevier Science: Amsterdam, The Netherlands, 2009.
4. Cavezza, F.; Boehm, M.; Terryn, H.; Hauffman, T. A Review on Adhesively Bonded Aluminium Joints in the Automotive Industry. *Metals* **2020**, *10*, 730. [CrossRef]
5. Pana, G.M. Developments of audi space frame technology for automotive body aluminum construction. *Appl. Mech. Mat.* **2020**, *896*, 127–132. [CrossRef]
6. Association, E.A. *EAA Aluminium Automotive Manual—Joining, The Aluminium Automotive Manual*; European Aluminium Association: Ljubljana, Slovenia, 2015; pp. 1–5.
7. Mercier, D.; Rouchaud, J.-C.; Barthés-Labrousse, M.-G. Interaction of amines with native aluminium oxide layers in non-aqueous environment: Application to the understanding of the formation of epoxy-amine/metal Interphases. *Appl. Surf. Sci.* **2008**, *254*, 6495–6503. [CrossRef]
8. Morsch, S.; Liu, Y.; Malanin, M.; Formanek, P.; Eichhorn, K.-J. Exploring wheter a buried nanoscale interphase exists within epoxy-amine coatings: Implications for adhesion, fracture thoughness and corrosion resistance. *Appl. Nano Mater.* **2019**, *2*, 2494–2502. [CrossRef]
9. Wu, G.; Dash, K.; Galano, M.L.; O'Reilly, K.A.Q. Oxidation studies of Al alloys: Part II Al-Mg alloy. *Corros. Sci.* **2019**, *155*, 97–108. [CrossRef]
10. Bechtel, S.; Nies, C.; Fug, F.; Grandthyll, S.; Müller, F.; Possart, W. Thin epoxy layers on native aluminium oxide: Specific ageing processes in mild conditions. *Thin. Solid Films* **2020**, *695*, 137756. [CrossRef]
11. Cornette, P.; Zanna, S.; Seyeux, A.; Costa, D.; Marcus, P. The native oxide film on a model aluminium-copper alloy studied by XPS and ToF-SIMS. *Corros. Sci.* **2020**, *174*, 108837. [CrossRef]
12. Takano, T.; Matsuya, H.; Kowalski, D.; Kitano, S.; Aoki, Y.; Habazaki, H. Raman and glow discharge optical emission spectroscopy studies on structure and anion incorporation properties of a hydrated alumina film on aluminum. *Appl. Surf. Sci.* **2022**, *592*, 153321. [CrossRef]
13. Strohmeier, B.R. An ESCA method for determining the oxide thickness in aluminum alloys. *Surf. Interface Anal.* **1990**, *15*, 51–56. [CrossRef]
14. Suárez-Campos, G.; Cabrera-German, D.; Castelo-González, A.O.; Avila-Avendano, C.; Ríos, J.L.F.; Quevedo-López, M.A.; Aceves, R.; Hu, H.; Sotelo-Lerma, M. Characterization of aluminum oxide thin films obtained by chemical solution deposition and annealing for metal-insulator-metal dielectric capacitor applications. *Appl. Surf. Sci.* **2020**, *513*, 145879. [CrossRef]
15. Din, R.U.; Gudla, V.C.; Jellesen, M.S.; Ambat, R. Accelerated growth of oxide film on aluminium alloys under steam: Part I: Effects of alloy chemistry and steam vapour pressure on microstructure. *Surf. Coat. Tech.* **2015**, *276*, 77–88. [CrossRef]

16. Evangelisti, F.; Stiefel, M.; Guseva, O.; Nia, R.P.; Hauert, R.; Hack, E.; Jeurgens, L.P.H.; Ambrosio, F.; Pasquarello, A.; Schmutz, P.; et al. Electronic and structural characterization of barrier-type amorphous aluminium oxide. *Electrochim. Acta* **2017**, *224*, 503–516. [CrossRef]
17. Liu, E.; Pan, Q.; Liu, B.; Ye, J. Microstructure Evolution of the near-surface deformed layer and corrosion behaviour of hot rolled AA7050 aluminium alloy. *Materials* **2023**, *16*, 4632. [CrossRef] [PubMed]
18. Lunder, O. Chromate-Free Pre-Treatment of Aluminium for Adhesive Bonding. Ph.D. Thesis, Norwegian University of Science and Technology, Trondheim, Norway, 2003.
19. Huttunen-Saarivirta, E.; Heino, H.; Vaajoki, A.; Hakala, T.J.; Ronkainen, H. Wear of additively manufactured tool steel in contact with aluminium alloy. *Wear* **2019**, *202934*, 432–433. [CrossRef]
20. Andreatta, F.; Turco, A.; de Graeve, I.; Terryn, H.; de Wit, J.H.W.; Fedrizzi, L. SKPFM and SEM study of the deposition mechanism of Zr/Ti based pre-treatment on AA6016 aluminum alloy. *Surf. Coat. Tech.* **2007**, *201*, 7668–7685. [CrossRef]
21. Duchoslav, J.; Kehrer, M.; Truglas, T.; Groiß, H.; Nadlinger, M.; Hader-Kregl, L.; Riener, C.K.; Arndt, M.; Stellnberger, K.H.; Luckeneder, G.; et al. The effect of plasma treatment on the surface chemistry and structure of ZnMgAl coatings. *Appl. Surf. Sci.* **2020**, *504*, 144457. [CrossRef]
22. Duchoslav, J.; Arndt, M.; Steinberger, R.; Keppert, T.; Luckeneder, G.; Stellnberger, K.H.; Hagler, J.; Riener, C.K.; Angeli, G.; Stifter, D. Nanoscopic view on the initial stages of corrosion of hot dip galvanized Zn-Mg-Al coatings. *Corros. Sci.* **2014**, *83*, 327–334. [CrossRef]
23. Arndt, M.; Duchoslav, J.; Itani, H.; Hesser, G.; Riener, C.K.; Angeli, G.; Preis, K.; Stifter, D.; Hingerl, K. Nanoscale analysis of surface oxides on ZnMgAl hot-dip-coated steel sheets. *Anal. Bioanal. Chem.* **2012**, *403*, 651–661. [CrossRef]
24. Tanuma, S.; Powell, C.J.; Penn, D.R. Calculations of electron inelastic mean free paths for 31 materials. *Surf. Interface Anal.* **1988**, *11*, 577–589. [CrossRef]
25. Rullik, L.; Evertsson, J.; Johansson, N.; Bertram, F.; Nilsson, J.-O.; Zakharov, A.A.; Mikkelsen, A.; Lundgren, E. Surface oxide development on aluminum alloy 6063 during heat treatment. *Surf. Int. Anal.* **2019**, *51*, 1214–1224. [CrossRef]
26. Cabrera, N.; Mott, N.F. Theory of the oxidation of metals. *Rep. Prog. Phys.* **1949**, *12*, 163–184. [CrossRef]

Disclaimer/Publisher's Note: The statements, opinions and data contained in all publications are solely those of the individual author(s) and contributor(s) and not of MDPI and/or the editor(s). MDPI and/or the editor(s) disclaim responsibility for any injury to people or property resulting from any ideas, methods, instructions or products referred to in the content.

MDPI
St. Alban-Anlage 66
4052 Basel
Switzerland
www.mdpi.com

Metals Editorial Office
E-mail: metals@mdpi.com
www.mdpi.com/journal/metals

Disclaimer/Publisher's Note: The statements, opinions and data contained in all publications are solely those of the individual author(s) and contributor(s) and not of MDPI and/or the editor(s). MDPI and/or the editor(s) disclaim responsibility for any injury to people or property resulting from any ideas, methods, instructions or products referred to in the content.